DIRE PREDICTIONS

UNDERSTANDING CLIMATE CHANGE
2ND EDITION

MICHAEL E. MANN
LEE R. KUMP

Penguin Random House

SENIOR EDITOR Peter Frances
PROJECT EDITOR Martyn Page
DESIGNERS Paul Drislane, Shahid Mahmood
MANAGING EDITOR Angeles Gavira Guerrero
ASSOCIATE MANAGING EDITOR Allison Singer
MANAGING ART EDITOR Michael Duffy
SENIOR PRE-PRODUCTION PRODUCER Luca Frassinetti
PRODUCER Vivienne Yong
ART DIRECTOR Karen Self
ASSOCIATE PUBLISHING DIRECTOR Liz Wheeler
PUBLISHING DIRECTOR Jonathan Metcalf

PEARSON

SENIOR EDITOR Christian Botting
PROGRAM MANAGER Anton Yakovlev
PROJECT MANAGER Crissy Dudonis
DEVELOPMENT EDITOR Karen Gulliver
DIRECTOR OF DEVELOPMENT Jennifer Hart
EXECUTIVE MARKETING MANAGER Neena Bali
EDITORIAL ASSISTANT Amy De Genaro

Second American edition, 2015
Published in the United States by DK Publishing
4th floor, 345 Hudson Street, New York, NY 10014
A Penguin Random House Company

15 16 17 18 19 10 9 8 7 6 5 4 3 2 1

01—274651—April/2015

A catalog record for this book is available from the Library of Congress

ISBN 978-1-4654-3364-0

MIX
Paper from
responsible sources
FSC™ C018179

The papers used for the pages and the cover are FSC certified, and come
from North America, where the book was printed. The inks used throughout
are vegetable inks and the special finish on the cover is biodegradable.

The future holds great promise for reducing our impact on Earth's
environment, and Pearson is proud to be leading the way. We strive to
publish the best books with the most up-to-date and accurate content, and
to do so in ways that minimize our impact on Earth. To learn more about
our initiatives, please visit www.pearson.com/responsibility.

Printed and bound by RR Donnelley, USA

www.dk.com

Contents

Part 1
CLIMATE
CHANGE BASICS

DIRE PREDICTIONS
UNDERSTANDING CLIMATE CHANGE

Part 2
CLIMATE CHANGE PROJECTIONS

Part 3
THE IMPACTS OF CLIMATE CHANGE

Part 4
VULNERABILITY AND ADAPTATION TO CLIMATE CHANGE

Part 5
SOLVING CLIMATE CHANGE

Introduction

Perhaps no topic in modern public discourse is so important—yet so poorly understood—as human-caused climate change. Our planet continues to warm, and a vast array of changes in Earth's climate are taking place as we continue to elevate the levels of greenhouse gases in the atmosphere through fossil fuel burning and other activities. If we continue on this course, we imperil our food and water supplies, the stability of our coastlines, the strength of our economy, our national security, and the health of our global environment. There is still time to avert catastrophic climate change, but the window of opportunity is beginning to close.

The Intergovernmental Panel on Climate Change (IPCC) was established in 1988 to evaluate the risk of human-caused climate change, and its periodic assessment reports have become the authoritative source for accurate information. These reports, however, are relatively impenetrable to the public.

In this book, esteemed climate scientists Michael E. Mann (who, along with other IPCC report authors, shared in the 2007 Nobel Peace Prize) and Lee R. Kump have partnered with the "information architects" at DK publishing and Pearson to produce *Dire Predictions*—essential reading for citizens of a world threatened

Storm over Las Vegas
A severe storm passes over Las Vegas airport in November 2012. Climate change may make powerful storms and other extreme weather more common.

by climate change. *Dire Predictions* presents and expands upon the findings documented in the IPCC 5th Assessment Report in an illustrated, visually stunning, and undeniably powerful way for the non-scientist.

In this second edition, Mann and Kump provide a comprehensive update on the state of our changing planet as assessed in the IPCC 5th Assessment Report. In this thoroughly updated book, they discuss new concerns, such as hidden tipping points in the climate system, a brief lull in the progressive warming of the planet, the news that 2014 was the warmest year on record, the acidification of the ocean, and loss of oxygen from the ocean's coastal regions and interior. These and many other relevant topics and events are explained—along with glimmers of hope arising from an emphasis on sustainability.

This edition also provides integrated media links that help illustrate some core science concepts, and the entire book is now available in ebook formats. The goal of this revision is to inform readers by clearly explaining important concepts about climate change—and to empower them to make informed decisions about actions that might make a difference.

The IPCC

The Intergovernmental Panel on Climate Change (IPCC) was established in 1988 by the United Nations Environment Program (UNEP) and the World Meteorological Organization (WMO). The panel was tasked with preparing a scientifically-based report on all relevant aspects of climate change and its impacts, and formulating possible strategies for addressing these impacts. The stated role of the IPCC is to "assess on a comprehensive, objective, open and transparent basis the scientific, technical and socioeconomic information relevant to understanding the scientific basis of risk of human-induced climate change, its potential impacts and options for adaptation and mitigation." The IPCC strives to be policy-relevant, but not policy-prescriptive.

Since its inception, the IPCC has reviewed and assessed the most recent scientific, technical, and socioeconomic information on climate change at regular intervals, periodically producing a set of comprehensive, well-documented reports. The IPCC reports summarize our continually improving knowledge of the underlying science of climate and convey the most reliable available projections for future climate change and its impacts. The reports are written by thousands of the world's leading scientists. Rigorous peer review is a hallmark of the IPCC process, and expert reviewers are called upon to comment on all aspects of the reports.

IPCC 5th Assessment Report

http://goo.gl/Vv6sLi

IPCC REPORTS

The information in *Dire Predictions* closely follows the findings of the IPCC Fifth Assessment Report. The authors have presented this material in a way that makes it accessible to non-scientists, and have supplemented the assessment's findings with additional and updated material.

About the authors

Dr. Michael E. Mann is Distinguished Professor of Meteorology at Penn State University. He is also director of the Penn State Earth System Science Center (ESSC).

Dr. Mann received his undergraduate degrees in Physics and Applied Math from the University of California at Berkeley, an

M.S. degree in Physics from Yale University, and a Ph.D. in Geology and Geophysics from Yale University. His research involves the use of theoretical models and observational data to better understand Earth's climate system. Honors and awards include NOAA's outstanding publication award in 2002 and selection by *Scientific American* as one of the fifty leading visionaries in science and technology in 2002. Dr. Mann shared, with other IPCC authors, in the award of the 2007 Nobel Peace Prize to the IPCC. He was awarded the Hans Oeschger Medal of the European Geosciences Union in 2012 and was awarded the National Conservation Achievement Award for science by the National Wildlife Federation in 2013. He made *Bloomberg News'* list of fifty most influential people in 2013. In 2014, he was named Highly Cited Researcher by the Institute for Scientific Information (ISI) and received the Friend of the Planet Award from the National Center for Science Education. He is a Fellow of both the American Geophysical Union and the American Meteorological Society.

Dr. Mann is author of more than 180 peer-reviewed and edited publications. He is author of *The Hockey Stick and the Climate Wars: Dispatches from the Front Lines* and a co-founder of the award-winning science website RealClimate.org.

Dr. Lee R. Kump is Professor and Head of the Department of Geosciences at Pennsylvania State University, and an associate of the Penn State Earth and Environmental Systems Institute (EESI), Earth System Science Center (ESSC), and the Penn State Astrobiology Research Center.

Dr. Kump received his bachelor's degree in geophysical sciences from the University of Chicago and his Ph.D. in Marine Science from the University of South Florida. He is a fellow of the Geological Society of America, American Geophysical Union, Geochemical Society, European Association of Geochemistry, the Canadian Institute for Advanced Research, and the Geological Society of London, and a Distinguished Alumnus of the Univeristy of South Florida. In his research he uses a variety of tools, including geochemical analysis and computer modeling, to investigate climate and biospheric change during periods of extreme and abrupt environmental and biodiversity change in Earth's history.

Dr. Kump is an active researcher with over 100 peer-reviewed and edited publications. His research has been featured in documentaries produced by *National Geographic*, the British Broadcasting Corporation (BBC), *NOVA Science-Now*, and the Australian Broadcasting Corporation. He is the lead author on the preeminent textbook in Earth System Science, *The Earth System*, and co-author of *Mathematical Modeling of Earth's Dynamical Systems*.

What's up with the weather (and the climate!)?

Climate change and global warming have been in the headlines for years. To truly understand these terms, and to appreciate how and why human activity is causing Earth's climate to change, you need first to understand what climate is, how it differs from weather, what factors affect it, and how modern human activity is altering it. The purpose of this section of the book is to introduce you to these concepts.

Climate and weather and us

We plan our daily activities around the weather. Will it rain? Is a storm or a cold front approaching? Weather is highly variable, and, although considerable improvements in weather forecasting have been made, it is still often unpredictable. Climate, on the other hand, varies more slowly and is highly predictable. We know what to expect of our local climate and the climate of other familiar regions. Panama, for example, is persistently warm and wet. Residents of Siberia and northern Alaska expect long and bitterly cold winters. In the mid-latitudes, a summer day is almost certainly going to be warmer than a winter day. Climate represents the average of many years' worth of weather. This averaging process smoothes out the individual blips caused by droughts and floods, tornadoes and hurricanes, and blizzards and downpours, while emphasizing the more typical patterns of temperature highs and lows and precipitation amounts.

The reason that climate is so predictable is that it is dependent on relatively fixed features of Earth, including Earth's spherical form, the shape of its orbit around the Sun, and its tilted axis of rotation. Other factors that determine climate have to do with the fact that Earth possesses both oceans and continents and a multi-layered atmosphere composed of various gases, including, critically, the **greenhouse gases** (▶ p.14).

Climate and latitude

Radiation from the Sun plays a big role in Earth's climate. The amount of radiation Earth receives from the Sun depends fundamentally on latitude. At the equator, the Sun's rays are most directly overhead and most directly focused on Earth's surface. As we move poleward, the Sun's position at noon is lower in the sky, and so its energy is spread over a larger area,

In the far north energy from the Sun is dispersed.

In the tropics energy from the Sun is concentrated.

Tropics face Sun all year round

South faces Sun in southern hemisphere

North faces Sun in northern hemisphere

Tilted axis

Climatic bands

- Polar regions
- Temperate zones
- The tropics

Earth-Sun Relations

http://goo.gl/KjZ8BH

making its energy less intense. This is the fundamental reason that the tropics (the region between the tropic of Cancer at 23½ °N and the tropic of Capricorn at 23½ °S) are warm and the poles are so cold.

Another factor that changes with latitude is seasonal contrast: how hot the summers are, and how cold the winters are. In the tropics, the difference in temperatures between summer and winter is fairly subtle, whereas at mid- to high-latitudes, the difference is significant. However, the existence of the seasons themselves depends not on latitude per se, but on the fact that Earth's spin axis, the imaginary

line that runs from pole to pole through the center of Earth, is tilted. Summer occurs in either hemisphere when the spin axis is inclined toward the Sun, while winter occurs when it is tilted away. The impact of spin axis tilt is most pronounced above the Arctic and Antarctic circles. This is why in these regions the Sun shines 24 hours a day during the summer while they remain in perpetual darkness during the winter.

Climate and the oceans

Another important factor determining continental climate is proximity to oceans. Water has a tremendous capacity for storing heat, much greater than that of the land. The oceans warm slowly during the summer and cool slowly during the winter, so coastal regions benefit from their moderating influence. In contrast, the continental interiors respond quickly to seasonal changes. This is why places like North Dakota and Saskatchewan typically have warm summers and cold winters compared to coastal locations.

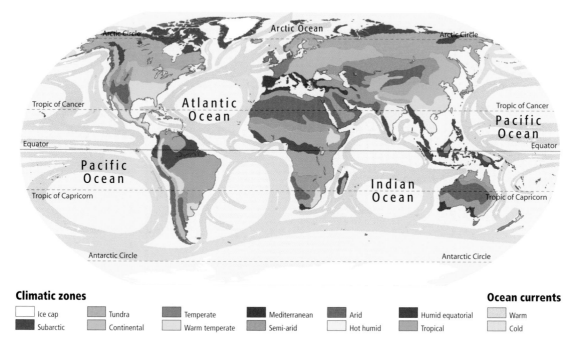

Climatic zones

- Ice cap
- Subarctic
- Tundra
- Continental
- Temperate
- Warm temperate
- Mediterranean
- Semi-arid
- Arid
- Hot humid
- Humid equatorial
- Tropical

Ocean currents

- Warm
- Cold

11

Climate and the atmosphere

The existence and composition of Earth's atmosphere also influences the climate. The **atmosphere** is the gaseous envelope surrounding Earth, and it is held in place by Earth's gravitational pull. Our atmosphere features distinct layers. The first 64–80 km (40–50 miles) above the surface contains 99% of the total mass of the atmosphere and is generally uniform in gas composition with some notable exceptions, including large variations in water vapor and high concentrations of **ozone**, known as the ozone layer, at 19 to 50 km (12 to 30 miles).

Atmospheric composition

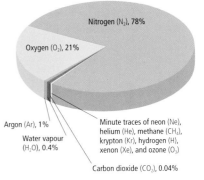

Nitrogen (N$_2$), 78%

Oxygen (O$_2$), 21%

Argon (Ar), 1%

Water vapour (H$_2$O), 0.4%

Minute traces of neon (Ne), helium (He), methane (CH$_4$), krypton (Kr), hydrogen (H), xenon (Xe), and ozone (O$_3$)

Carbon dioxide (CO$_2$), 0.04%

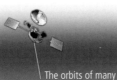

The orbits of many satellites lie within the exosphere.

Exosphere
Outermost layer of the atmosphere. Extends to about 10,000 km (6,000 miles).

Thermosphere
Extends to about 640 km (400 miles).

Mesosphere
Rises from about 50 to 80 km (30 to 50 miles) above the surface. Air becomes cooler as the altitude increases.

Stratosphere
Extends upward to a height of about 50 km (30 miles). Contains atmospheric ozone layer. Temperature increases with altitude through the stratosphere, inhibiting vertical air currents, and making the stratosphere highly stable, in contrast to the troposphere.

Troposphere
Layer in contact with Earth's surface. Extends from the surface to about 8 to 17 km (5 to 10 miles). Air temperature decreases with altitude, leading to instability. Less dense air sits below more dense air, which results in air movements and storm generation. "Weather" takes place almost exclusively within the troposphere.

Sea level

Mesopause
Boundary between the mesosphere and the thermosphere.

Stratopause
Boundary between the stratosphere and the mesosphere.

Atmospheric ozone layer
Layer within the stratosphere. Absorbs ultraviolet solar radiation, so warming the surrounding atmosphere.

Tropopause
Boundary between the troposphere and the stratosphere.

OZONE LAYER

Radiation can be beneficial. Our planet would be cold without the Sun's rays, and plants need **solar radiation** to photosynthesize. But radiation can also be dangerous, particularly ultraviolet (UV) radiation. Fortunately, oxygen and ozone molecules in the **stratosphere** absorb most of the UV radiation reaching Earth. Ozone is a compound of oxygen that contains three atoms (the oxygen gas we breathe contains two oxygen atoms) and it is a lung irritant and smog producer in surface air pollution (in the **troposphere**). However, in the stratosphere, ozone protects life on Earth by absorbing UV radiation. In the process, the radiation destroys the chemical bonds between the oxygen atoms in the ozone molecule. Under normal circumstances, the ozone molecules rapidly reform. Unfortunately we've polluted the atmosphere with **chlorofluorocarbons**, and these molecules accelerate the destruction of ozone, reducing its abundance and thus its ability to protect us from UV radiation.

Atmospheric circulation

To understand rainfall patterns, a major player in climate, we need to understand the basic principles of atmospheric circulation.

The pattern of rising moist air near the equator and sinking dry air in the subtropics is referred to as the "Hadley Circulation." The Hadley Circulation is a key component of the general circulation of the atmosphere; it helps to transport heat from the equatorial region to higher latitudes. Because of the Hadley Circulation, generally the tropics are warm and wet, while the subtropics are warm and dry. And as a result of the atmospheric circulation patterns found at higher latitudes, the mid-latitude regions experience large seasonal contrasts in temperature and rainfall patterns, while the polar regions are generally cold and dry. Rainfall in the mid-latitudes is related to the "polar front." Those of us who live in North America or Europe may know the polar front by a different name—the "storm track"—an expression that refers to the day-to-day variations in the location and intensity of the polar front.

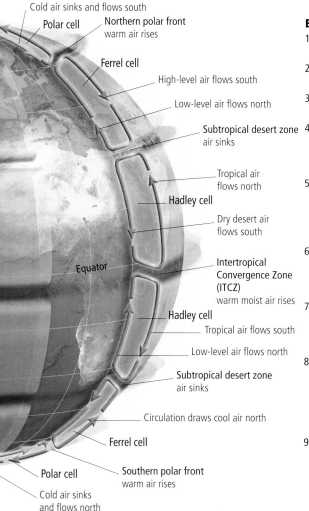

Cold air sinks and flows south
Polar cell
Northern polar front
warm air rises
Ferrel cell
High-level air flows south
Low-level air flows north
Subtropical desert zone
air sinks
Tropical air flows north
Hadley cell
Dry desert air flows south
Equator
Intertropical Convergence Zone (ITCZ)
warm moist air rises
Hadley cell
Tropical air flows south
Low-level air flows north
Subtropical desert zone
air sinks
Circulation draws cool air north
Ferrel cell
Polar cell
Southern polar front
warm air rises
Cold air sinks and flows north

Basic principles of atmospheric circulation

1. Water evaporates from land and ocean and becomes water vapor, a gas that composes part of the lower atmosphere.

2. Like a huge hot-air balloon, air near the ground in the tropics warms as a result of solar radiation, becomes buoyant, and rises.

3. As the warm tropical air rises it expands, and, like gas coming out of a spray can, it cools.

4. Cold air holds less water vapor, so as rising tropical air cools, water condenses out as droplets that congeal and form towering cumulus clouds and rainfall-producing thunderstorms. Thus, the tropics are rainy.

5. The rising air in the tropics draws surface air in from higher latitudes, forming the Intertropical Convergence Zone (ITCZ). The ITCZ migrates north and south within the tropics as seasons change. The ITCZ is associated with trade winds that converge near the equator.

6. Air rising in the tropics moves poleward once it reaches higher altitudes. Because Earth is spinning, this poleward flow gets disrupted, and air sinks at approximately 30°S and 30°N (the subtropics).

7. This air sinking in the subtropics is now quite dry because most of the water vapor was precipitated out of it when the air was rising. Furthermore, as air descends it gets compressed and warms. This is why deserts tend to occur at subtropical latitudes.

8. A second region of rising air exists in middle-to-high latitudes (roughly 40–60°N and 40–60°S) in the region known as the polar front. Here, surface air from lower latitudes encounters cold polar air heading toward the equator. The denser polar air forces itself underneath the warmer air mass, causing it to rise, cool, and condense out its water vapor.

9. Finally, air near the poles sinks, causing the polar regions to be arid.

Global Atmospheric Circulation

http://goo.gl/FYo4Ld

Climate change and us

Looking back on Earth's history, it comes as no surprise that climates change. Indeed, on any timescale—centennial, millennial, or over millions of years—the climate record is anything but constant. Over the last two million years, as ice sheets advanced and retreated across northern North America and Scandinavia, climates have oscillated between very cold and more pleasant, like today. Geologists have designated this interval of time an **ice age** and divide it into two epochs: the Pleistocene, which lasted from two million years ago until 10,000 years ago, and the Holocene, which encompasses the last 10,000 years. Ice ages are marked by episodes of extensive **glaciation**, alternating with episodes of relative warmth. The colder periods are called **glacials**, the warmer periods are referred to as **interglacials**. The Holocene is the most recent interglacial period of Earth's most recent ice age.

Prior to the Holocene epoch, just 20,000 years ago, the world was gripped by a glacial climate with ice sheets covering much of North America and Scandinavia. Prior to two million years ago there were no large ice sheets in the northern hemisphere, and prior to 34 million years ago there were no large ice sheets anywhere. Throughout the time of the dinosaurs (the Mesozoic Era, 252–65 million years ago) the world was considerably warmer than today, and during the warmest intervals, reptiles and other cold-intolerant organisms lived above the Arctic circle. We have to go back into **deep time**, more than 300 million years ago, to find evidence for a previous ice age, one that included glacial periods perhaps considerably colder and more extensive than the most recent Pleistocene glacial epoch.

Despite this backdrop of natural climate fluctuations on various timescales, human greenhouse gas emissions (see box below), are creating an atmosphere unlike anything Earth has experienced for tens of millions of years. In many respects there may be no geologic precedent for the accelerated rate of the disturbance we are imposing on Earth's climate

GREENHOUSE GASES AND EARTH'S CLIMATE

The greenhouse effect occurs on our planet because the atmosphere (the gaseous cloud that surrounds Earth) contains greenhouse gases. Greenhouse gases are special because they absorb heat, warming the atmosphere around them. Greenhouse gases exist naturally in Earth's atmosphere in the form of water vapor, carbon dioxide, methane, and other trace gases, but atmospheric concentrations of some greenhouse gases, such as carbon dioxide and methane, are being increased as a result of human activity. This increase occurs primarily as a result of the burning of **fossil fuels**, but also through deforestation and agricultural practices. Certain greenhouse gases, such as CFCs, and the surface ozone found in smog (which is distinct from the natural ozone found in the lower stratosphere), are produced exclusively by human activity.

 Carbon dioxide (CO_2)

 Water vapor (H_2O)

 Ozone (O_3)

 Methane (CH_4)

 Nitrous oxide (N_2O)

ICE KINGDOMS

Scientists refer to the cold regions of the planet where water persists in its frozen form (i.e., regions covered with glaciers and ice sheets or with permanently frozen soils) as the cryosphere. Much of the cryosphere exists near the poles, but high-altitude mountain glaciers occur at lower latitudes. **Glaciers** are huge masses of ice formed from compacted snow. An **ice sheet** is a mass of glacier ice that covers surrounding terrain and is greater than 50,000 square kilometers (20,000 square miles). The only current ice sheets are in Antarctica and Greenland.

The two most important regions of the **cryosphere** are the continental ice sheets of Antarctica and Greenland. These huge expanses of glacial ice significantly affect the amount of solar energy reflected to space, but their most important role is their storage of water. If the ice sheets were to melt completely, sea level would rise by about 80 m (260 ft). Much of this storage is in the East Antarctic ice sheet, which is less likely to be affected by anthropogenic warming in the next few centuries; West Antarctica and Greenland melting would cause a more modest, but nevertheless devastating, 12 m (39 ft) of sea-level rise. In contrast, the expansion and contraction of sea ice (floating ice near the poles) has no effect on sea level, but can dramatically affect ocean circulation, local climate, and **ecosystems**. Perennially frozen ground (permafrost) influences soil water content and vegetation over vast regions and is one of the cryosphere components most sensitive to atmospheric warming trends (▶ p.148). Other regions of the cryosphere are also responding to climate change: the seasonal minimum sea-ice coverage of the Arctic Ocean is currently diminishing (▶ p.148), and most mountain glaciers are shrinking (▶ p.64).

A Tour of the Cryosphere

http://goo.gl/Bjd57G

Southern hemisphere

South pole · Antarctica · Antarctic circle · 66.5°S

Northern hemisphere

North pole · Greenland · Arctic circle · 66.5°N

system; the resulting impacts may be quite unlike those associated with past natural climate variation.

How do we, as individual citizens, best address this problem? A key first step is becoming informed about the nature of the threat, and the potential solutions that are available. We hope that this book will equip readers with the information necessary to make wise choices—because it is becoming increasingly clear that the decisions we make impact our collective future world.

The edge of the Greenland ice sheet

The Greenland and Antarctic ice sheets have largely survived the glacial/interglacial fluctuations of the last two million years, whereas the North American (Laurentide) and Scandinavian (Fennoscandian) ice sheets have come and gone.

Part 1
Climate Change Basics

Basic principles of physics and chemistry dictate that Earth will warm as concentrations of greenhouse gases increase. Though various natural factors can influence Earth's climate, only the increase in greenhouse gas concentrations linked to human activity, principally the burning of fossil fuels, can explain recent patterns of global warming. Other changes in Earth's climate, such as shifting precipitation patterns, worsening drought in many locations, increasingly severe heat waves, and more intense Atlantic hurricanes, are also likely repercussions of human impact.

Key ideas

Greenhouse gases

- Our climate changes over time, influenced by both human and natural causes.

- Gases in the atmosphere called greenhouse gases absorb heat and warm the atmosphere around them.

- Changes to the atmosphere caused by greenhouse gases are modified by positive and negative feedback loops.

Human-caused change

- Three greenhouse gases have been rising at dramatic rates for the last two centuries, driven by fossil-fuel burning, deforestation, and agriculture.

- Over a similar timescale, Earth's surface has been getting warmer.

- Over many millions of years, fluctuations in surface temperatures have been mirrored by changes in the amount of carbon dioxide in the atmosphere.

- Human-caused climate change may also be linked to the spread of drought conditions and the increased frequency of heat waves and the potential destructiveness of hurricanes.

Modeling climate change

- Earth's climate can be reproduced in computer models. The predictions of these models provide a good fit with actual observations and suggest that recent warming cannot be explained by natural factors alone.

The relative impact of humans and nature on climate

A variety of human actions, as well as natural factors, can potentially affect Earth's climate.

Natural impacts
Natural factors influencing climate include:

- **The Sun.** Over time, small but measurable changes occur in the output of the Sun—Earth's ultimate source of warming energy.

- **Volcanic eruptions.** Explosive volcanic eruptions modify the composition of the atmosphere by injecting small particles called **aerosols** into the atmospheric layer known as the stratosphere (◄ p.12), where they may reside for several years. These particles either reflect or absorb incoming solar radiation that would otherwise warm Earth's surface.

- **Earth's orbit.** While changes in Earth's orbit relative to the Sun (▶ p.67) influence climate on timescales of many millennia, they are not thought to play any significant role on the shorter timescales relevant to modern climate change.

Mount Pinatubo
The 1991 Mount Pinatubo eruption in the Philippines was the most explosive volcanic eruption of the 20th century. It had a cooling effect on Earth's surface for several years after the eruption.

The Changing
Face of Earth

http://goo.gl/i6nWxt

Human impacts

The main human impact on climate is an enhanced greenhouse effect (▶ p.24), leading to a warming of the lower atmosphere (the **troposphere**; ◀ p.12). This is caused by increased atmospheric concentrations of greenhouse gases, primarily carbon dioxide produced by fossil-fuel burning (▶ p.26). There are also several secondary impacts of human activity. One of these is the introduction of aerosols like those ejected into the atmosphere by volcanoes. These small particles (mostly sulfate and nitrate) are suspended in the atmosphere by industrial activity, such as coal combustion. Industrial aerosols reside in the lower atmosphere for only a short amount of time, and therefore must constantly be produced in order to have a sustained climate impact. The impacts of aerosols are more regionally limited and more

variable than those of the well-mixed greenhouse gases. Aerosols generally reflect solar radiation back into space, creating a regional cooling influence. However, certain aerosols (including black carbonaceous aerosols) can, like greenhouse gases, have a surface warming influence instead.

Other human impacts include ozone depletion in the stratosphere and changes in land use such as tropical deforestation, which modifies the absorptive and energy-exchange properties of Earth's surface.

Why is climate changing?

Because of the different ways these factors influence the pattern of solar radiation reaching Earth, scientists can distinguish which factors are most likely responsible for any given observed change in climate (▶ p.78). Indeed, scientists have now determined that while natural factors have been responsible for substantial changes in climate in past centuries and millennia, human impacts, particularly increased greenhouse gas concentrations, appear to be responsible for the major climate changes of recent decades (▶ p.32).

Industrial pollutants
Factories not only emit greenhouse gases such as carbon dioxide into the atmosphere, they also produce significant amounts of tiny particles (aerosols) that can affect the climate.

Human impacts appear to be responsible for the major climate changes of recent decades.

Taking action in the face of uncertainty

The role of scientists in global policy making

Uncertainty in climate change projections exists whether we like it or not. Some people express skepticism in response to this uncertainty, and cite it as an excuse for inaction. Scientists themselves are trained to be skeptical. They recognize that few things in science can be stated with certainty, that hypotheses can only be disproved, not proved, and that results and conclusions should be expressed in terms of this uncertainty. While scientists can make strong conclusions from uncertain results, others view uncertainty as an indicator of flawed or inadequate scientific approaches.

Why are climate projections uncertain?

In climate science, uncertainty arises from a variety of sources, including the inherently unpredictable nature of aspects of the physical climate system and of the human factors driving climate change; the necessary simplifications that occur when computer models are created; and incomplete knowledge about critical parameters in these models. In determining the likelihood that a conclusion is correct, climate scientists often turn to statistics, but some factors cannot be quantified by data. In these cases, likelihoods can only be established based on expert judgment.

POSSIBLE PATHS OF FUTURE GLOBAL WARMING

Here we see the path warming has taken in the recent past (solid orange line with the uncertainty shown with the blue lines above and below) and a range of possible projected outcomes through the year 2050, based on 300 individual model simulations using various models and various assumptions concerning fossil-fuel use, mitigation strategies, demographic and economic patterns, and assuming no future large volcanic eruptions. The green area includes 90% of the outcomes, whereas the outer green lines encompass all the simulation results. The amount of warming is therefore relative to the average global temperature from 1986–2005 (left scale) or 1850–1900 (right scale).

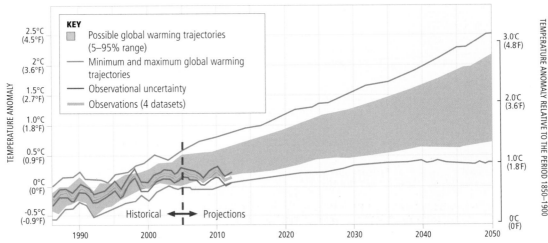

TEMPERATURE ANOMALY

KEY
- Possible global warming trajectories (5–95% range)
- Minimum and maximum global warming trajectories
- Observational uncertainty
- Observations (4 datasets)

2.5°C (4.5°F)
2°C (3.6°F)
1.5°C (2.7°F)
1.0°C (1.8°F)
0.5°C (0.9°F)
0°C (0°F)
-0.5°C (-0.9°F)

Historical ← → Projections

1990 2000 2010 2020 2030 2040 2050

TEMPERATURE ANOMALY RELATIVE TO THE PERIOD 1850–1900

3.0°C (4.8F)
2.0°C (3.6F)
1.0°C (1.8F)
0°C (0F)

Scientific conclusions arise from time-tested theories, accurate observations, realistic models based on the fundamentals of physics and chemistry, and consensus among colleagues working in the discipline.

The Fifth Assessment Report

The Fifth Assessment Report of the IPCC presents conclusions in terms of the likelihood of particular outcomes. These are expressed as a probability, based on the quality, volume, and consistency of the evidence and the extent of agreement among experts. Likelihood ranges from virtually certain (more than 99% probability of occurrence) to exceptionally unlikely (less than 1% probability of occurrence).

As policymakers are aware, the risk associated with any of these projections is the combination of the *probability of occurrence* and the *severity of the damage* if it were to occur. This means that we should not ignore projections labeled as "unlikely." If any of those events actually occurred, the consequences would be dire.

IPCC 5th
Assessment Report

http://goo.gl/Vv6sLi

IPCC PROJECTIONS FOR THE LATE 21ST CENTURY

This table outlines the IPCC's projections for the late 21st century, ranked in decreasing order of certainty.

VIRTUALLY CERTAIN (99–100%)
- Cold days and nights will be warmer and less frequent over most land areas
- Hot days and nights will be warmer and more frequent over most land areas
- The extent of permafrost will decline
- Ocean acidification will incresae as the atmosphere accumulates CO_2
- Northern hemisphere glaciation will not initiate before the year 3000
- Global mean sea level will rise and continue to do so for many centuries

VERY LIKELY (90–100%)
- Arctic sea ice cover will continue to shrink and thin, and northern hemisphere spring snow cover will decrease
- The dissolved oxygen content of the ocean will decrease by a few percent
- The rate of increase in atmospheric CO_2, methane, and nitrous oxide will reach levels unprecedented in the last 10,000 years
- The frequency of warm spells and heat waves will increase
- The frequency of heavy precipitation events will increase
- Precipitation amounts will increase in high latitudes
- The ocean's conveyor-belt circulation will weaken
- The rate of sea level rise will exceed that of the late 20th century
- Extreme high sea-level events will increase, as will ocean wave heights of mid-latitude storms

LIKELY (66–100%)
- If the atmospheric CO_2 level stabilizes at double the present level, global temperatures will rise by between 1.5°C (2.7°F) and 4.5°C (8.1°F)
- Areas affected by drought will increase
- Precipitation amounts will decline in the subtropics
- The loss of glaciers will accelerate in the next few decades
- Climate change will promote ozone-hole expansion, despite an overall decline in ozone-destroying chemicals

ABOUT AS LIKELY AS NOT (33–66%)
- Intense tropical cyclone activity will increase
- The West Antarctic ice sheet will pass the melting point if global warming exceeds 5°C (9°F)—this is relative not absolute

UNLIKELY (0–33%)
- Antarctic and Greenland ice sheets will collapse due to surface warming

VERY UNLIKELY (0–10%)
- The oceans' conveyor-belt circulation will shut down abruptly
- Methane from seafloor clathrates will be released catastrophically

EXCEPTIONALLY UNLIKELY (0–1%)
- If the atmospheric CO_2 level stabilizes at double the present level, global temperatures will rise by <1.0°C (1.8°F) (exceptionally unlikely) or by >6.0°C (10.8°F) (very unlikely)

0 10 20 30 40 50 60 70 80 90
PROBABILITY (%)

Why is it called the greenhouse effect?

While the label has stuck, the greenhouse effect in our atmosphere is not exactly like an actual greenhouse. A greenhouse lets in solar radiation (mostly in the form of visible light), which keeps it warm and allows the plants inside to grow. The greenhouse stays warm primarily because its glass windows prevent the wind from carrying away the heat. This is very different from the greenhouse effect.

The greenhouse effect occurs on our planet because the atmosphere (the gaseous cloud that surrounds Earth) contains greenhouse gases. Greenhouse gases are special—they absorb heat, which then warms the atmosphere around them. Not all gases are greenhouse gases; nitrogen and oxygen—the most abundant gases in the atmosphere—aren't greenhouse gases. Life on Earth requires some atmospheric warming to exist, provided by greenhouse gases including water vapor, carbon dioxide, and methane. Without its greenhouse atmosphere, Earth's temperature would plummet to well below freezing.

We know that Earth has been a habitable planet for over 3 billion years. This means that there has been a strong greenhouse effect for at least this long. The carbon dioxide that human activity has added to the atmosphere isn't creating the greenhouse effect—it's simply intensifying it.

Hot house
The greenhouse effect does keep the planet warm like the plants inside this greenhouse, but it functions somewhat differently than a real greenhouse.

HOW THE GREENHOUSE EFFECT WORKS

Greenhouse gases allow solar radiation to pass through the atmosphere and heat Earth, but they interfere with the loss of heat from the land and ocean, redirecting some of that heat back to the surface.

1 Earth absorbs solar radiation and warms up.

Sun

Earth

Earth's atmosphere (not to scale)

Solar radiation

Solar radiation heats Earth

2 Like all warm objects, Earth begins to radiate heat.

Earth radiates heat, which is absorbed by its atmosphere

3 Heat radiating from Earth encounters greenhouse gas molecules in the atmosphere, and is absorbed. The atmosphere warms; as a result, it too radiates heat. Some of this heat is radiated out into space, but the rest is radiated back to Earth's surface. This extra energy warms Earth to higher temperatures. When averaged over several years, the energy radiated into space very nearly balances the solar energy absorbed by Earth. Currently, however, Earth is radiating slightly less heat into space than it is receiving from the Sun, because of the recent addition of greenhouse gases to the atmosphere. Consequently, the planet is warming.

Atmosphere radiates heat into space

Atmosphere radiates heat back to Earth

Earth's surface radiates more heat to the atmosphere than it is receiving from the Sun because of greenhouse gases

Atmosphere & Energy Balance

http://goo.gl/SqZlle

Positive feedback loops compound the greenhouse effect of carbon dioxide

When we think about the effects of adding carbon dioxide and other greenhouse gases to the atmosphere, we have to think not just about what these gases might do to climate directly, but also about their indirect effects. This is particularly important when addressing the criticism leveled by global warming skeptics that climate researchers over-emphasize the effects of carbon dioxide (CO_2), but ignore the fact that water vapor is the largest contributor to the greenhouse effect. Changes in water vapor content, however, occur primarily in response to warming or cooling caused by changes in atmospheric CO_2, solar heating, or other direct effects. Indirect effects, such as water vapor fluctuations, are often the result of what are known as **feedback loops**.

Clouds from both sides
Clouds can be involved in both negative and positive climate feedback loops. Low clouds tend to cool the planet, whereas high clouds, such as the cirrus ones shown here, warm the planet.

Direct radiative effect

Climate scientists refer to the "direct radiative effect" of CO_2 and other greenhouse gases—that is, the effect that a particular gas has on the energy budget of the planet. In terms of direct radiative effect, CO_2 is important, but water vapor is an even larger contributor to the overall greenhouse effect. Knowing how much indirect warming greenhouse gas emissions will cause is trickier, because of the complex feedback loops that are set in motion when greenhouse gases are added to the atmosphere.

POSITIVE FEEDBACK LOOP

○ Adding CO_2 to the atmosphere tends to warm it, causing global warming.

● Since water vapor is a greenhouse gas, the atmosphere tends to warm even more as water vapor increases.

● The warm atmosphere causes more surface water to evaporate and become water vapor.

Increasing CO_2 → Global warming → Increased water vapor

increases evaporation

warms atmosphere

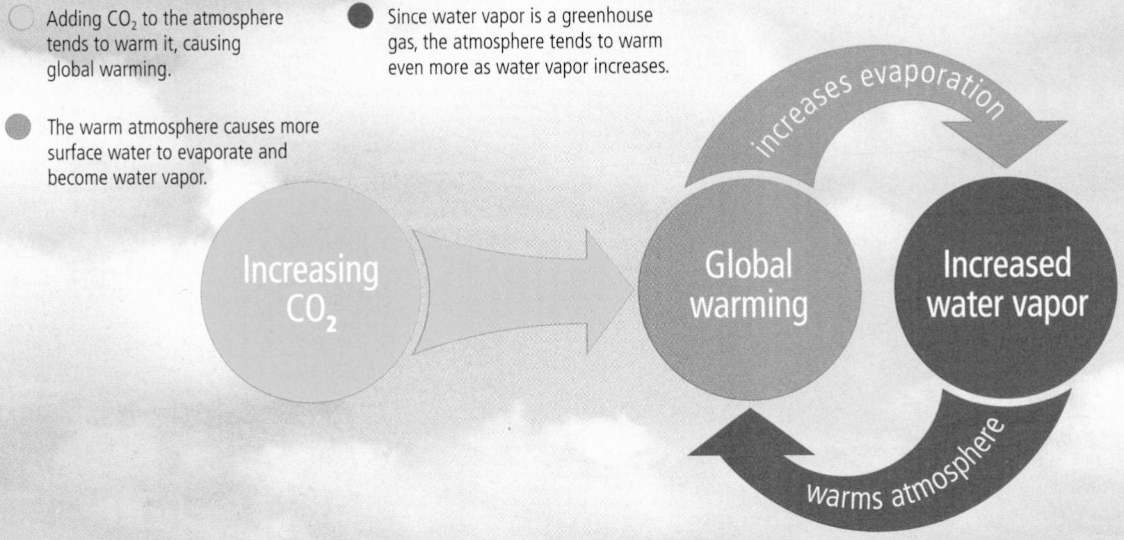

Positive feedback

Adding CO_2 to the atmosphere tends to warm the atmosphere. The initial warming causes more surface water to evaporate, increasing the atmospheric water vapor content. Since water vapor is a greenhouse gas, the atmosphere will then tend to warm even more. This effect, known as the "water vapor feedback loop," is a positive feedback loop, because it amplifies the original change. Similarly, a modest amount of warming at high latitudes (in Alaska and Scandinavia, for example) can lead to a substantial melting of snow and ice, exposing the soil, rocks, or the ocean surface. Because these surfaces are less reflective than snow, they absorb more solar radiation, thereby warming even more rapidly. This effect constitutes another very important positive feedback loop.

Negative feedback

Some of the additional water vapor in the atmosphere will condense to form clouds. Clouds contribute to the greenhouse effect by trapping heat in the atmosphere, but they also reflect solar energy back to space, helping to cool the planet. Depending on where the clouds form, their overall effect therefore may be to either cool or warm the atmosphere. So things become even more complicated. If low clouds become more prevalent in response to increased CO_2, they have a cooling effect, thus offsetting some of the initial warming. This is a negative feedback loop, as it reduces the original change.

Observations and modeling demonstrate that the overall effect of clouds is to cool the planet; this means that the warming induced by the buildup of CO_2 is likely to be somewhat less than it would be if the only role that water vapor played was that of an additional greenhouse gas. However, positive feedbacks in the climate system outweigh negative feedbacks, so the expected warming from CO_2 buildup is greater than its direct radiative effect alone.

NEGATIVE FEEDBACK LOOP

○ Adding CO_2 to the atmosphere tends to warm it, causing global warming.

● The warm atmosphere causes more surface water to evaporate and become water vapor.

● Some water vapor condenses to form clouds. Clouds contribute to the greenhouse effect by trapping heat in the atmosphere, but they also reflect solar energy back to space. On balance, reflection dominates—so clouds tend to cool the planet.

Increasing CO_2

Global warming

Increased low clouds

increases evaporation

cools atmosphere

What are the important greenhouse gases, and where do they come from?

Although carbon dioxide (CO_2) has been the primary focus of concern in human-induced climate change, there are a number of other anthropogenic (human-generated) gases that also affect the radiation balance of the planet. Most of these—CO_2, methane (CH_4), nitrous oxide (N_2O) and others—aren't exclusively anthropogenic; with the exception of the CFCs (chlorofluorocarbons), they exist naturally. In fact, some of these gases are produced and consumed by natural processes at tremendous rates.

The carbon cycle

Consider the CO_2 produced by your great-grandparents when they lit their coal stove several decades ago. After being dormant in the coal for perhaps hundreds of millions of years, the carbon atoms were exposed to a high temperature in the stove, causing them to react with oxygen to produce CO_2.

Let's follow a single CO_2 molecule through the **carbon cycle** to learn more. The CO_2 molecule in the stove escaped out the chimney into the atmosphere, where it was taken on a whirlwind tour of the planet.

CARBON RECYCLING

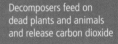

Green plants take in carbon dioxide during photosynthesis

Carbon dioxide gas in air

Decomposers feed on dead plants and animals and release carbon dioxide

Rotting plants and animals return carbon to the soil

Burning coal and other fossil fuels release carbon dioxide into the air

Sometime during its first decade of travels, the molecule entered the interior of a leaf via photosynthesis, where its two oxygen atoms were stripped away, and its carbon atom became part of the leaf. At the end of the season, the leaf fell to the forest floor. Bacteria or fungi consumed the leaf, reattaching two oxygen atoms to the

Rice paddies
Rice paddies, such as these in the North Alps of Japan, are major methane emitters because their flooded soils provide an ideal habitat for bacteria that produce methane as a metabolic byproduct.

carbon, and the resulting new CO_2 molecule was released back into the atmosphere. In the several decades since the carbon atom was freed from its lump of coal, it could have been part of five different plants. This indicates that the lifetime of a CO_2 molecule in the atmosphere is about one decade. However, this cycle of uptake and release is balanced; it doesn't remove carbon dioxide, it just recycles it. Only processes acting much more slowly (over hundreds or thousands of years), including stirring the gas into the ocean, provide a net removal mechanism of CO_2.

Release without uptake

While CO_2 has always been released into the atmosphere by natural processes, fossil-fuel burning and deforestation are relatively new sources of atmospheric CO_2. Since this input hasn't been matched by increased removal, the result has been a continuous rise in atmospheric CO_2 over the last 200 years. Even if we stopped burning fossil fuels today, the return to pre-industrial levels of atmospheric CO_2 would take several centuries.

Bacterial byproduct

Another contributor to the human greenhouse amplification is methane. Methane is a natural gas, as well as an anthropogenic one. It is a metabolic byproduct of the microbes that inhabit oxygen-poor environments, such as the black mud of ponds and rice paddies, and the guts of cattle and termites. Because of the low availability of oxygen, a gas they cannot tolerate, these microbes consume organic matter, but produce CH_4 rather than CO_2 as a byproduct. While some bacteria consume CH_4, most CH_4 released into the atmosphere is removed by chemical reactions that yield CO_2. Thus an increase in CH_4 leads to an increase in CO_2. The average atmospheric lifetime for a CH_4 molecule is about a decade. Agriculture, principally rice cultivation and livestock release, together with industrial and landfill sources, have increased the rate of CH_4 production in recent decades, and atmospheric levels have risen correspondingly.

Agriculture is also the source of N_2O, another potent greenhouse gas. N_2O is the natural byproduct of microbes in soils and the ocean, but anthropogenic sources include nitrogen fertilizer, tropical deforestation, and the burning of fossil fuels. These human sources have increased the flow of N_2O into the atmosphere by 40–50% over pre-industrial levels; consequently, the N_2O content of the atmosphere has risen steadily. In the atmosphere, N_2O is slowly broken down by sunlight; the average time an N_2O molecule spends in the atmosphere is a little over a century.

Keeping up
with Carbon

http://goo.gl/pDpuk0

(Cont.)

》》
(Cont.)

Refrigerants cause global warming

Freons, including the **chlorofluorocarbons (CFCs)**, were once seen as a wonder gas, because they were efficient, nontoxic refrigerants. Only in the late 20th century did scientists realize that these gases were involved in the destruction of the ozone layer, and many are now banned. The ozone layer is a region of the stratosphere rich in ozone gas, which protects life on Earth from ultraviolet radiation. Unfortunately, CFCs are also greenhouse gases. Unlike the other greenhouse gases, CFCs, their replacements the hydrofluorocarbons and hydrochlorofluorocarbons, and other similar gases used as refrigerants and fire extinguishers, have no natural source.

Global warming potentials

The capacity for the various greenhouse gases to cause climate change differs because each molecule interacts with heat differently. To compare the various gases, researchers have introduced the concept of **global warming potential (GWP)**. A gas's GWP is a calculation of the increase in greenhouse effect caused by the release of a kilogram of the gas, relative to that produced by an equivalent amount of CO_2. GWPs have to be expressed in terms of a time horizon, such as 20, 100, or 500 years after release of the gas, because the different greenhouse gases have different atmospheric lifetimes. The table opposite illustrates the strong greenhouse capabilities of methane, nitrous oxide, and freons. Fortunately, they are being emitted at much slower rates than CO_2, so their overall effect is still less than that of CO_2. Since CO_2 is involved in so many processes, it has a range of lifetimes. However, remember that these processes recycle CO_2 rather than remove it from the atmosphere for good. Its ultimate lifetime, therefore, is considerably longer than that of the other greenhouse gases. As time passes, the relative importance of CO_2 will only increase.

AMOUNT OF GAS IN THE 2013 ATMOSPHERE EXPRESSED AS PARTS PER BILLION (PPB)

CO$_2$ (carbon dioxide)
Amount in atmosphere: 395,000 ppb

CH$_4$ (methane)
Amount in atmosphere: 1,800 ppb

N$_2$O (nitrous oxide)
Amount in atmosphere: 325 ppb

CFC-11 (trichlorofluoromethane)
Amount in atmosphere: 0.235 ppb

HFC-134a (1,1,1,2-tetrafluoroethane)
Amount in atmosphere: 0.070 ppb

CF$_4$ (carbon tetrafluoride)
Amount in atmosphere: 0.080 ppb

LIFETIME AND GLOBAL WARMING POTENTIAL OF ANTHROPOGENIC GREENHOUSE GASES

Gas	CO$_2$	CH$_4$	N$_2$O	CFC-11	HFC-134a	CF$_4$
Lifetime years	Multiple	12	121	45	13	50,000
Global warming potential of a pulse of this greenhouse gas compared to CO$_2$						
After 20 years	1	86	268	7,020	3,790	4,950
After 100 years	1	34	298	5,350	1,550	7,350
After 500 years	1	8	153	1,620	435	11,200

Consider the simultaneous release of a kilogram of carbon dioxide and methane. The atmospheric lifetime of methane (CH$_4$) is 12 years. In the short term (the first 20 years after the gas is released), methane is a strong greenhouse gas, 86 times more powerful than CO_2. However, because it has a shorter lifetime than CO_2, on century-long timescales it becomes only 34 times as potent, and after 500 years, its potency has been significantly reduced.

Scrapped appliances
Although the production of CFC-11 and CFC-12 has been banned by international agreement because they also destroy the ozone layer, these gases still leak out of automobiles and from air conditioners and refrigerators decomposing in landfills and contribute to global warming (and ozone depletion).

Greenhouse gases on the rise

How do we know the composition of ancient air?

Although scientists have only been measuring the amount of greenhouse gas in the atmosphere for the last few decades, nature has been collecting samples for hundreds of thousands of years.

Over millennia, as snow accumulated on the Antarctic and Greenland ice sheets, the pressure of overlying snow has compressed buried layers of snow into ice. Air trapped in the snow became encapsulated in tiny bubbles. Scientists can drill into those ice sheets, remove samples called ice cores, extract the gas from the bubbles trapped in the ice, and measure the composition of ancient air.

Tragically, the ice-core archive of ancient atmospheres is melting away as climates warm.

Scientists in Antarctica drill into the ice sheet.

An ice core is removed from the drill barrel.

Ancient air
A close-up of a cross-section slice of an ice core clearly shows the tiny bubbles of gas that were trapped when the ice formed.

Distinguishing layers
Ash or dust may show up as dark bands in ice cores. These bands can provide information on wind speeds, desertification, and volcanic eruptions.

The impact of human activity

Together with modern observations, analyses of these ice cores reveal the unambiguous human effect on atmospheric composition. As the graphs on the right demonstrate, three greenhouse gases— carbon dioxide (CO_2), methane (CH_4), and nitrous oxide (N_2O)—have been rising at dramatic rates for the last two centuries. Driven by fossil-fuel burning, deforestation, and agriculture, the recent skyrocketing trends greatly exceed the natural fluctuations of the preceding hundreds of thousands of years. CO_2 has increased by 40%, CH_4 by 150%, and N_2O by 20%. These gases have a powerful effect on climate, despite the fact that their concentrations are measured in parts per million (ppm) or billion (ppb). You might have to sort through millions of atmospheric molecules to find one of these molecules.

The Greenland ice sheet
An aerial view of the Greenland ice sheet near Baffin Bay, taken in April 2013. The second largest body of ice in the world (after the Antarctic ice sheet), the Greenland ice sheet contains tiny bubbles of ancient air, analysis of which can give information about past atmospheric composition.

CHANGES IN GREENHOUSE GASES: ICE-CORE AND MODERN DATA

Atmospheric concentrations of carbon dioxide, methane, and nitrous oxide are shown here for the last 10,000 years. Concentrations have increased dramatically since the Industrial Revolution.

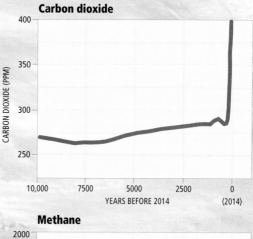

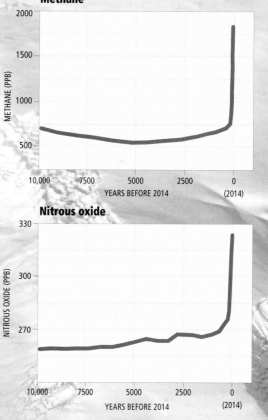

Is the increase in atmospheric CO_2 the result of natural cycles?

Some talk-radio hosts and other global warming skeptics have claimed that the undeniable rise in atmospheric carbon dioxide (CO_2) levels over the last 50 years could simply be a natural fluctuation.

How do scientists know that it is not?

There are several clues that convince scientists that the CO_2 increase is due almost entirely to fossil-fuel burning.

1 Because fossil-fuel consumption is such an integral part of the global economy, utilization rates are reasonably well known. Looking at these numbers, scientists can determine that fossil-fuel burning can more than account for the recent rise in atmospheric CO_2. In fact, the recent CO_2 rise equates to only half of what has actually been released into the atmosphere. Scientists had to probe deeper to find out what has happened to the other half: it has dissolved into the ocean or been taken up by the growth of forests.

No natural source for CO_2 buildup has been identified.

2 Earth's atmosphere is naturally radioactive because carbon-14 (radiocarbon) forms in the upper atmosphere. From measurements of tree-ring radioactivity (an indicator of atmospheric radioactivity), we know that this radioactivity was relatively high prior to the Industrial Revolution. However, it decreased over the first half of the 20th century while CO_2 levels were increasing (until nuclear bomb tests reversed the trend in radiocarbon). This indicates that much of the additional carbon driving the rise in atmospheric CO_2 levels is coming from a low radioactivity or "radiocarbon dead" source. Volcanoes and the deep ocean are radiocarbon-dead sources, as are fossil fuels. But we know the source couldn't be volcanoes or the deep ocean, because…

3 Carbon atoms exist in three forms, or **isotopes**. Each has six protons, but each has a different number of neutrons in its nucleus. The most abundant form of carbon, representing nearly 99% of all carbon, is carbon-12, an atom that has six protons and six neutrons. Carbon-12 is stable (non-radioactive). A more rare form of stable carbon is carbon-13, with six protons but seven neutrons. When plants and algae photosynthesize, they preferentially use molecules of CO_2 that contain carbon-12. Thus when scientists analyze carbon sources that were derived from organic matter, like the fossil

fuels coal and oil, they find that the carbon has a low ratio of carbon-13.

Just as the atmosphere has gradually become less radioactive over time, its ratio of carbon-13 to carbon-12 has been decreasing at a rate that can only be explained by a carbon source that has very low carbon-13/carbon-12 ratios achieved exclusively through photosynthesis. This rules out natural, non-plant derived carbon sources, such as volcanoes and the oceans.

Isotopes unambiguous about fossil fuels

The combined trends in the atmosphere's radioactivity and its carbon-13/carbon-12 ratio are satisfactorily explained by only one source: fossil-fuel burning. While scientists acknowledge that uncertainties exist in our knowledge of global warming, the source of the carbon that has led to the recent buildup of atmospheric carbon dioxide isn't one of them.

Radioactive trees
Tree rings record annual growth increments, and the carbon in them records changes in atmospheric radioactivity over time.

WHAT THE NUMBERS TELL US...

The rise in atmospheric CO_2 since 1800 (graph a) is undeniable. It closely matches the increase in human-generated CO_2 emissions, which are quite well known (graph b). The radioactivity of the atmosphere has been decreasing (graph c), implying that the source of the increase is radiocarbon-dead. Also, the ratio of carbon-13 to carbon-12 has been decreasing, implying that the source of the increase was derived from organic matter or plants (graph d). All this points conclusively to fossil fuels as the main cause of the rise in atmospheric CO_2.

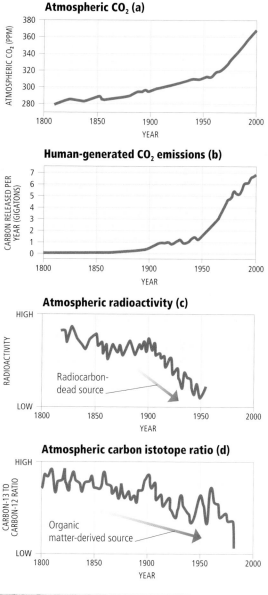

Atmospheric CO₂ (a)

Human-generated CO₂ emissions (b)

Atmospheric radioactivity (c)

Atmospheric carbon istotope ratio (d)

It's getting hotter!
Surface temperature observations

Thermometer records have been kept for more than a century across much of the globe; and during the last few decades, records have been kept almost worldwide. Records include surface air temperatures measured over continents and islands, and sea surface temperatures measured over oceans. By averaging these data across the globe, it is possible to estimate average global temperatures back to the mid-19th century, although uncertainties increase as we look back in time and the data become sparser. We rely on more limited instrumental or historical records, supplemented by indirect evidence, to deduce temperature changes prior to the mid-19th century (▶ p.48).

Accelerated warming

The instrumental temperature record shows that surface warming has taken place across the oceans and land, and that the rate of warming has accelerated over the most recent decades. The average rate of global warming over the full 20th century was slightly less than 0.10°C (0.18°F) per decade, but in the past few decades the warming rate has increased to more than 0.15°C (0.27°F) per decade. Overall, the average temperature of the globe has warmed from about 13.5°C (56.3°F) to 14.5°C (58.1°F) since the beginning of the 20th century.

While this warming of roughly 1.0°C (1.8°F) might seem small, it is nearly one-fourth of the estimated change in the temperature of the globe between today and the depths of the last ice age, when the area that is now New York City was covered by a sheet of ice almost half a kilometer (about

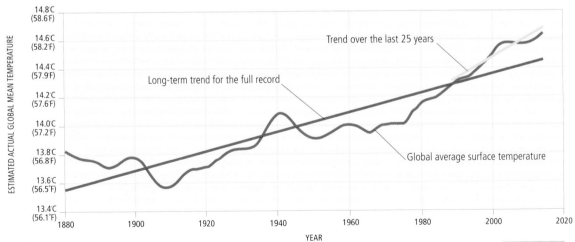

TRENDS IN GLOBAL AVERAGE SURFACE TEMPERATURE
Global temperature has risen just under 1.0°C (1.8°F) since 1880 (blue curve). The rate of warming in recent decades (yellow line) has increased by 50% relative to the longer-term trend (red line).

Taking Earth's Temperature

http://goo.gl/B3UbWC

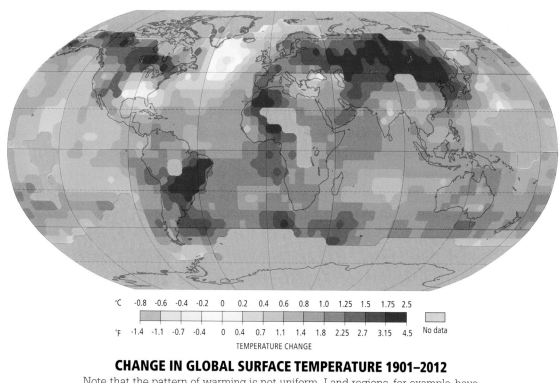

°C	-0.8	-0.6	-0.4	-0.2	0	0.2	0.4	0.6	0.8	1.0	1.25	1.5	1.75	2.5		No data
°F	-1.4	-1.1	-0.7	-0.4	0	0.4	0.7	1.1	1.4	1.8	2.25	2.7	3.15	4.5		

TEMPERATURE CHANGE

CHANGE IN GLOBAL SURFACE TEMPERATURE 1901–2012

Note that the pattern of warming is not uniform. Land regions, for example, have warmed more than the oceans.

a third of a mile) thick. The warming observed so far is only a small fraction of the total warming expected during the course of the next century, if we continue to burn fossil fuels at current rates (▶ p.98).

No urban heat bias

Can we trust what the instrumental temperature record is telling us? Some argue that there may be an "urban heat" bias in the record, due to the fact that cities have warmed up artificially because of their high rate of energy use and the dark, sunlight-absorbing properties of streets and blacktop. Since many records come from urban areas, some argue that this bias may contaminate estimates of global temperature trends.

Scientists, however, have accounted and corrected for these impacts in their assessments of global temperature trends. Furthermore, similar trends are seen when only rural measurements are used.

There are other data, too, with which to counter these misconceptions: thermometer measurements indicate that the ocean surface is warming significantly as well. Obviously there is no urbanization impact on sea surface temperatures.

Conclusion

Independent temperature data from the atmosphere (▶ p.38), the ground, and the ocean sub-surface (▶ p.36), combined with evidence such as melting snow, ice, and permafrost (▶ p.110), rising sea levels, and observed changes in plant and animal behavior make it clear that Earth's surface is warming noticeably.

Where is all that heat going?

While global warming technically refers to warming of Earth's surface, most of the heat contained within the climate system is actually stored at depth within the oceans. Not surprisingly, much of the warming due to increased greenhouse gas concentrations has gone into heating the oceans. A substantial amount of heat has penetrated down into the upper (first 700 m/2300 ft) and deep (below 700 m/2300 ft) levels of ocean.

Though the warming of Earth's surface has slowed slightly over the past decade due to natural factors (▶ pp.94–95), the oceans have accumulated heat quite rapidly over this same timeframe. Some of this increase in oceanic heat burial (heat that is "buried" below the ocean surface and diffuses down into the ocean waters) is associated with the El Niño phenomenon.

El Niño and La Niña

During **El Niño** events (▶ pp.100–101), the surface warming over a large part of the tropical Pacific Ocean allows more heat to escape from the ocean surface out into space. That heat loss comes at the expense of a decrease in the burial of heat beneath the ocean surface. Conversely, **La Niña** conditions—the flip side of El Niño—are associated with a cooler tropical Pacific Ocean surface, but increased burial of heat beneath the ocean surface.

The predominance of La Niña conditions in the tropical Pacific during the past decade thus means that the relatively cool conditions over a large part of Earth's tropical oceans surface has been balanced by accelerated penetration of heat down into the upper- and mid-depths of the ocean.

Global sea level rise

Ocean heat burial contributes to global sea level rise (▶ pp.110–111), because water expands as it warms. The steady rise in ocean heat content is thus adding to the effect of accelerating ice loss to steepen the rate of sea level rise.

Ocean warming
A school of blackfin barracuda *(Sphyraena qeni)*, viewed from below. Much of the warming due to climate change has increased the quantity of heat in the oceans. Ocean warming not only contributes to a rise in global sea levels but also affects marine life, ocean currents, and global weather patterns.

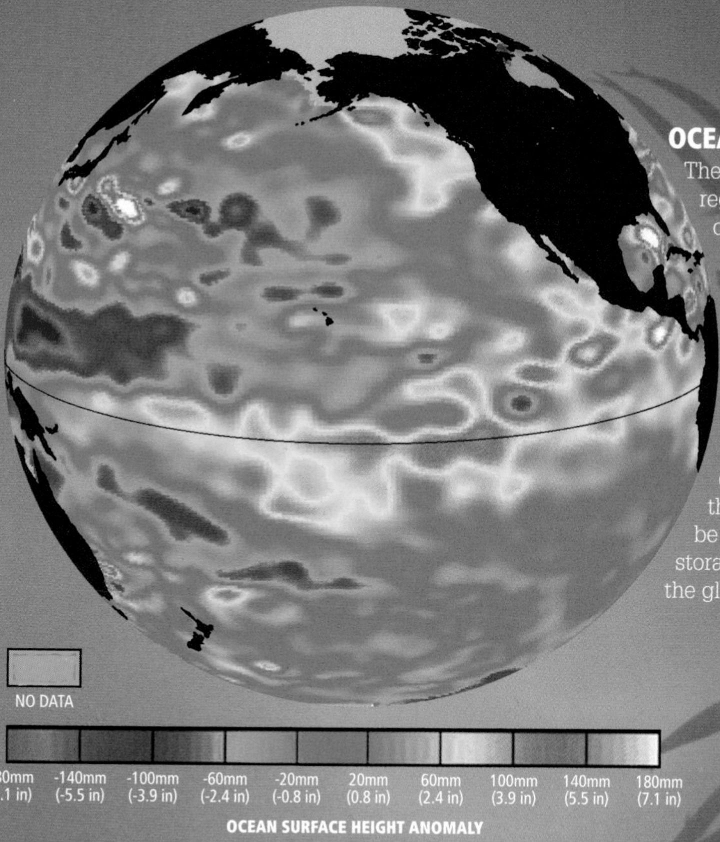

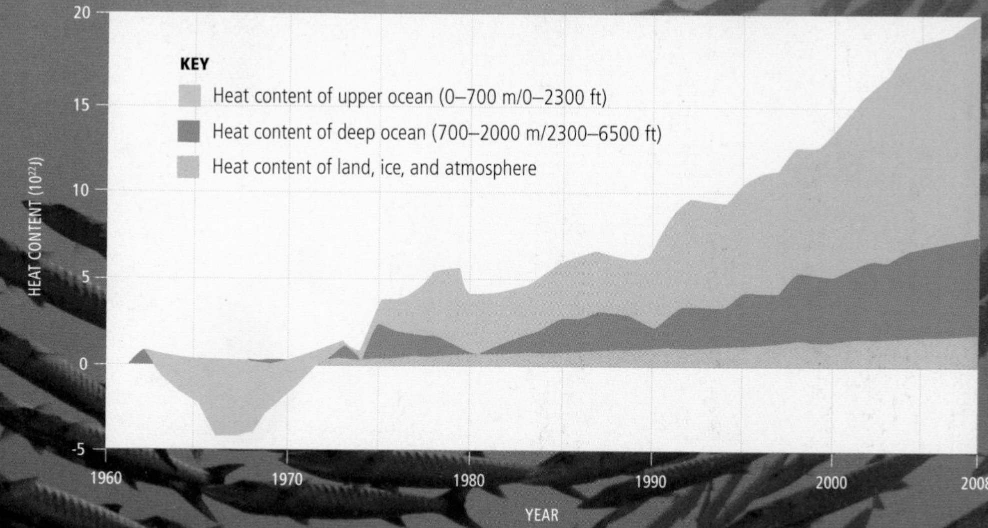

NO DATA
-180mm (-7.1 in)
-140mm (-5.5 in)
-100mm (-3.9 in)
-60mm (-2.4 in)
-20mm (-0.8 in)
20mm (0.8 in)
60mm (2.4 in)
100mm (3.9 in)
140mm (5.5 in)
180mm (7.1 in)
OCEAN SURFACE HEIGHT ANOMALY

OCEAN SURFACE HEIGHT ANOMALY

The height of the ocean surface varies regionally; warm water is less dense than cold water, so it is higher than the mean height of the ocean. This 2014 satellite image of the Pacific Ocean has been color coded to show anomalies in the height of the sea surface, with the red end of the spectrum indicating above-normal sea surface heights and the blue end showing below-normal heights. These height anomalies are differences from what is normal for the time of year and region. They can be used to investigate ocean heat storage and its possible influence on the global climate.

KEY
Heat content of upper ocean (0–700 m/0–2300 ft)
Heat content of deep ocean (700–2000 m/2300–6500 ft)
Heat content of land, ice, and atmosphere
HEAT CONTENT (10^{22}J)
20
15
10
5
0
-5
1960
1970
1980
1990
2000
2008
YEAR

ENERGY COMPARISON

This graph shows the amount of energy that has gone into increasing the heat content of the oceans compared to the amount of energy that has gone into increasing the heat content of other components of the climate system (land, ice, and atmosphere) during the past 50 years.

Is our atmosphere really warming?

At the time of the Third Assessment Report of the IPCC in 2001, there was a body of observational data that some skeptics claimed contradicted the evidence for global warming. Two measurement sources of atmospheric temperatures over the past few decades—one from **microwave** measurements made with satellites, the other from weather balloon data—seemed to show that the lower atmosphere (the troposphere) was warming only minimally. This contradicted ground-based thermometer measurements, which indicated substantial surface warming. Skeptics argued that the inconsistency showed either that the surface data were flawed and the warming trend they indicated was in error (◀ p.34), or that the surface warming was not caused by increased greenhouse gases, since models predicted that the troposphere would warm by as much or even more than the surface if greenhouse gases increase.

More recently, however, problems have been found in the older satellite and weather balloon-based assessments. The satellite estimates were compromised by errors that artificially converted some positive trends into negative trends. It turns out that the weather balloon data had not been sufficiently quality controlled to eliminate unreliable records.

Now that these problems have been corrected, there is considerably greater agreement between the various atmospheric temperature estimates. Corrected assessments of the satellite and weather balloon data indicate a cooling trend in recent decades in the lower stratosphere, and warming trends at the surface and in the troposphere above it. This is precisely the pattern of atmospheric temperature change predicted by climate model simulations of the response to increased greenhouse gas concentrations (▶ pp.80–81).

ATMOSPHERIC LAYERS

Exosphere
is the outermost layer of the atmosphere. It extends to about 10,000 km (6,000 miles)

The orbits of many satellites lie within the exosphere.

Mesopause
is the boundary between the mesosphere and the thermosphere

Stratopause
is the boundary between the stratosphere and the mesosphere

Ozone layer
(within the stratosphere) absorbs harmful ultraviolet radiation

Tropopause
is the boundary between the

ATMOSPHERIC TEMPERATURE TRENDS

These graphs show observed temperature trends at various altitudes in the atmosphere. (Temperatures represent departures from the 1961–1990 average.)

ATMOSPHERIC TEMPERATURE CHANGES

This graphic shows the pattern of 20th century (1890–1999) atmospheric temperature changes predicted by climate models. Note that the greatest warming is observed in the tropics and in the lower atmosphere.

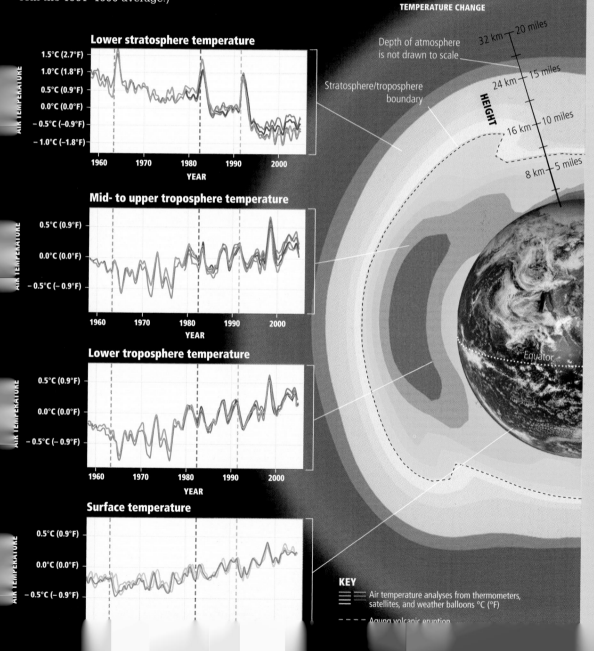

Back to the future

Deep time holds clues to climate change

When a doctor receives a new patient, a detailed health history is taken. Did the patient suffer previous ailments, and if so, what was the course of the illness? What are the current symptoms? What brought on the illness? A physician needs to know these things before making a diagnosis.

Like your body's temperature regulatory system, Earth's climate is self-regulatory, able to resist change, but subject to disturbances.

Global warming, in this context, is a planetary "fever"—not particularly high now, but possibly heading toward critical extremes.

If Earth has a planetary fever, then geologists are acting as "geo-physicians," compiling the patient's history by delving into the rock record. Sedimentary rocks preserve a record of climate history, so studying rocks can tell scientists if there is any link between changing levels of atmospheric CO_2 and climate over Earth's 4.6 billion–year history.

Paleoclimate reconstruction

Geologists are pursuing two lines of inquiry: What were the climates of the distant past, and what were the corresponding atmospheric CO_2 levels? By collecting data, geologists have established a fairly continuous record of ocean and atmospheric temperatures that spans tens of millions of years. As a result, past climate change is quite well understood. Over the last two million years—a time period referred to as the Pleistocene glacial epoch—climates have oscillated between very cold and more livable, like our current climate. By extracting air bubbles from ice cores (◄ p.30), scientists know that atmospheric CO_2 levels have varied with these temperature swings over the last several hundred thousand years.

The last 65 million years

On longer timescales, we find that climates were generally warmer than today; the last glacial era was over 300 million years ago. The record for the

Sediment study
Scientists spend months on the ocean-drilling vessel *JOIDES Resolution*, recovering sediment cores from the deep-ocean floor.

JOIDES Resolution

last 65 million years (since the extinction of the dinosaurs) is shown below. Note that the poles were considerably warmer in the past; in the Eocene optimum, for example, alligators and sequoia forests were thriving above the Arctic Circle. The gradual cooling over the past 50 million years is curious—was it caused by declining atmospheric CO_2 levels?

To extend the record of variations in atmospheric CO_2 levels, geologists have applied knowledge from other fields, including biology, biochemistry, and soil science, to develop "proxy" measures.

ESTIMATE OF PAST POLAR TEMPERATURE

Sediment cores show that polar temperatures 50 million years ago were up to 14°C (25°F) warmer than today, and have since fallen in a series of steps.

The central derrick houses the drill string—thousands of meters of pipe that support the drill bit on the sea floor below.

GEOLOGIC RECORDS OF CO_2 LEVELS

Estimates of atmospheric CO_2 levels, based on various proxies, show that levels have fallen over the last 44 million years.

(Cont.)

(Cont.)

Proxy measures

Geologic proxies are characteristics of sedimentary rocks that we can use to infer ancient environmental conditions. Climate scientists study proxies because there are no thermometer records or samples of ancient air from the geologic past. Proxies may even include the pores on fossil leaves: when atmospheric CO_2 levels are low, plants need more pores to bring in more CO_2. Under the microscope, well-preserved fossils reveal the number of pores; comparing this pore density to that of living plants allows scientists to establish CO_2 levels in the geologic past.

Fossilized imprint of ancient fern leaf

Leaf pore

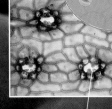

Leaf pores in ancient fern

When alive, this tree fern needed a lower density of pores because CO_2 levels were high.

Fossilized leaf pore

Living fern leaf

Leaf pore

Leaf pores in living fern

This modern leaf has a higher density of pores because the current atmospheric CO_2 levels are low.

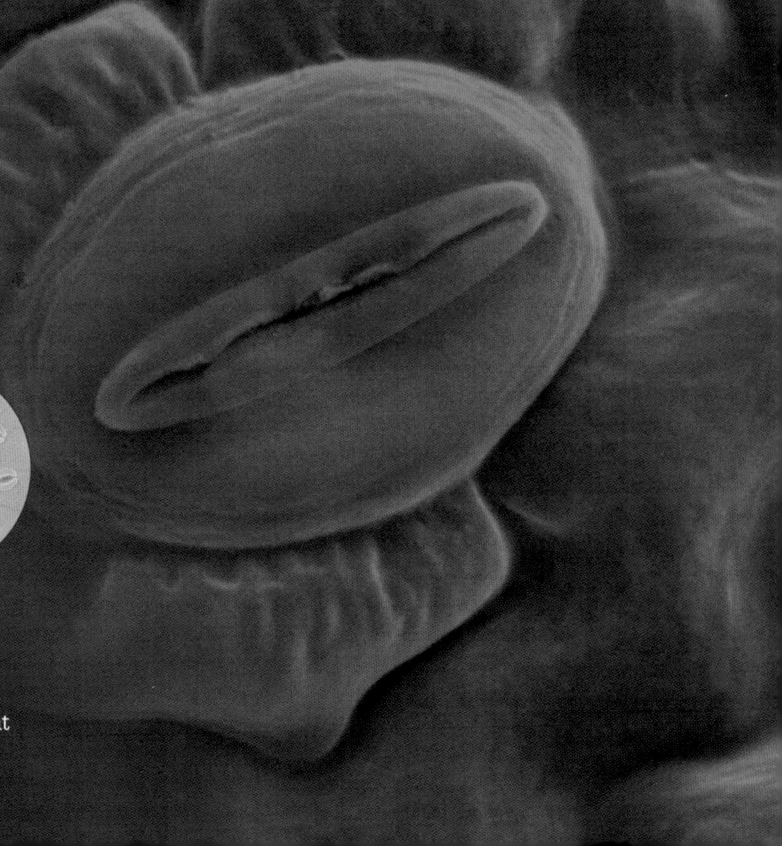

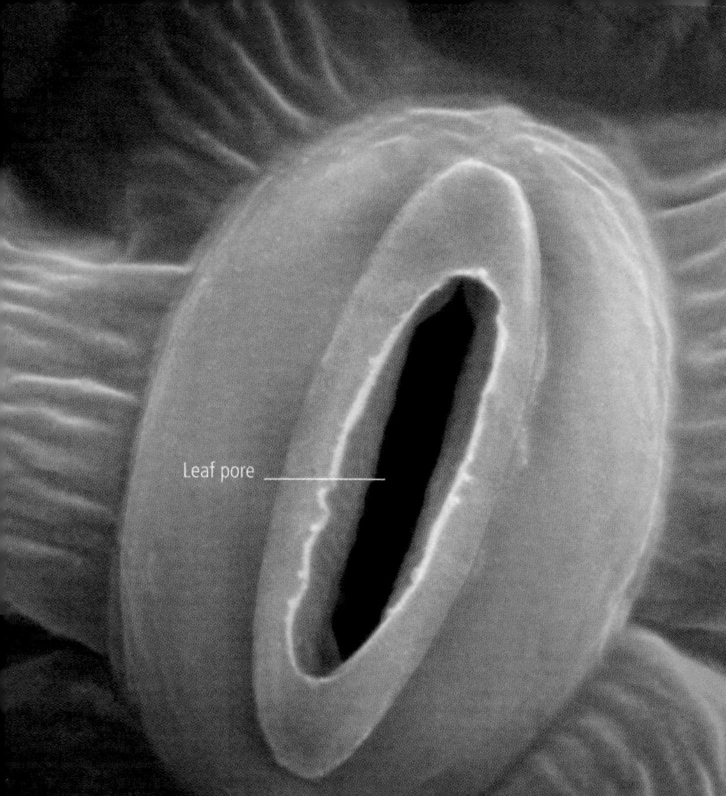

The function of leaf pores
The leaf pores (stomata) of a plant allow it to collect CO_2 from the atmosphere for photosynthesis. Few stomata on a leaf indicate that atmospheric CO_2 levels are relatively high; conversely, the presence of many stomata indicates low CO_2 levels.

Leaf pore

Interpreting the results

Taken together, the geologic records of climate and atmospheric CO_2 levels reveal an expected relationship: when carbon dioxide levels were high, the climate was warm; when CO_2 levels were low, the climate was cooler. The correlation isn't perfect, and the mismatches are areas of current research. In particular, the overall cooling trend from 15 million years ago to the present doesn't seem to be reflected in a reduction in atmospheric CO_2. Are the proxies for climate and CO_2 in error during these times, or are there other climate variables that we are neglecting? In this case, the growth of polar ice sheets and their high reflectivity may have provided extra cooling.

Modern measurements of atmospheric CO_2 are nearing 400 ppm. Current estimates of available fossil-fuel reserves translate into the potential for atmospheric CO_2 to rise above 2000 ppm.

If we use existing fossil-fuel reserves and do nothing to capture the CO_2 released, atmospheric CO_2 will exceed anything experienced on Earth for over 50 million years.

Plant Productivity in a Warming World

http://goo.gl/Yljkqr

Suffocating the ocean

Have you ever had a carbonated beverage go "flat"? If so, you probably know that if you had put it in the refrigerator, it would have stayed carbonated longer. That's because cold liquids hold more gas than warm liquids. The ocean is no different; the amount of oxygen dissolved from the atmosphere in the ocean's uppermost layer, where there is a constant exchange of gases with the atmosphere, depends on temperature. So it should come as no surprise that as the planet warms, the oceans are losing oxygen.

Effects of ocean warming

Ocean warming does two things. First, it reduces the amount of oxygen dissolved in the cold, dense polar oceans that sink to the deep sea, providing essential oxygen to organisms living in the ocean's depths.

Second, it reduces the tendency for this water to sink by making it less dense, further isolating the oxygen-depleted deep waters from the more oxygen-rich surface waters. In both cases, the ocean loses oxygen. Highly productive regions of the surface ocean become underlain by strongly oxygen-depleted subsurface waters, and if and when these rise to the surface, oxygen-dependent fish and other animals can be killed, creating oceanic "dead zones."

Ocean warming in Earth history

The geologic record confirms this connection between warming and deoxygenation. "Greenhouse" intervals of Earth history, when climates were warm, are notable for the frequency and global extent of "oceanic anoxic events" that led to mass extinction. Often triggered by

Fish die-off in Louisiana
Hundreds of thousands of dead fish carpet the water in Bay Jimmy, Louisiana, in September 2010. This huge fish die-off was likely caused by low-oxygen conditions that regularly occur in the Gulf of Mexico, although it may have been made worse by a recent oil spill in the Gulf. Dead fish included sting rays, menhaden, catfish, speckled trout, and redfish.

extended periods of tremendous volcanic eruptions that released huge quantities of carbon dioxide, these events were characterized by extreme global warmth, ocean stratification, and increased runoff from the land surface that brought nutrients into the ocean, stimulating biological productivity and increasing oxygen demand in the deep ocean, leading to anoxia. As the event unfolded, anoxia spread from greater ocean depths toward the surface, reducing the regions of the ocean that could serve as refuge for animals to critically few. Mass extinction ensued.

Modern ocean warming

Ocean dead zones have expanded over the last few decades, and human activity has been implicated. In addition to the effects of warming from the buildup of atmospheric carbon dioxide, increased agricultural runoff is fertilizing the oceans, making them more productive at the surface but choking off oxygen at depth. Earth history reveals the potential consequences.

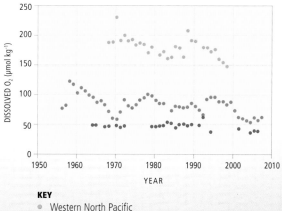

KEY
- Western North Pacific
- Eastern North Pacific
- Eastern Equatorial Pacific

OXYGEN CONTENT OF THE PACIFIC OCEAN

The oxygen content of subsurface waters in three regions of the Pacific Ocean (the Western North Pacific, Eastern North Pacific, and Eastern Equatorial Pacific) has been declining over the last several decades.

Weren't scientists warning us of a coming ice age only decades ago?

Hollywood license
The 2013 movie *Snowpiercer* is set in a frozen Earth brought about by a failed experiment to solve global warming through climate engineering.

A myth still common today in popular critiques of global warming science involves the supposed consensus among climate scientists in the 1970s that Earth was headed into an ice age. As this has obviously turned out not to be the case—the argument typically goes—why should we believe what scientists are saying today about global warming? As is typical with urban myths, there is a small grain of

> ## "Scientists ponder why world's climate is changing; a major cooling widely considered to be inevitable"
>
> *THE NEW YORK TIMES* MAY 21, 1975

truth to this claim. Ultimately, however, the assertion is incorrect and misleading on several grounds.

It is true that decades ago climate scientists were uncertain about future trends in global average temperature, but there was no consensus that an ice age was imminent, or even that the future trend would be one of cooling. Those ideas were conveyed in alarmist accounts in the popular media during the mid-1970s (e.g., in *Newsweek, Time*, and *The New York Times*)—not in scientific publications. In fact, the National Academy of Sciences concluded in a 1975 report: "…*we do not have a good quantitative understanding of our climate machine and what determines its course. Without the fundamental understanding, it does not seem possible to predict climate…*"

So why all the uncertainty? First, scientists recognized that Earth was eventually due for another ice age as part of the natural cycles of cold ("glacial") and warm ("interglacial") periods that occur due to slow changes in Earth's orbit around the Sun. When the next ice age might be was not well known. Second, some scientists thought that global warming could trigger a shut-down in the so-called "conveyor belt" circulation pattern that transports warm water to the higher latitudes of the North Atlantic, counter-intuitively causing cooling in Europe and North America. That scenario is now viewed as unlikely. Finally, and most significant of all, scientists were already aware that human impacts on climate included both a regional cooling effect from industrial aerosols and the global warming effect of increased greenhouse gas concentrations due to fossil-fuel burning. But it was still not fully understood which of these effects would dominate in the end. We know now that the cooling from the 1950s to the 1970s was probably due to a substantial increase in the regional aerosol cooling impact, which at the time was overwhelming the greenhouse warming impact (▶ p.72).

The rate of increase in greenhouse gas concentrations due to fossil-fuel burning has accelerated, while policies such as the Clean Air Acts have dramatically reduced aerosol production in most industrial regions. So the impact of increasing greenhouse gas concentrations has considerably overtaken any aerosol-related cooling. As a result, since the 1970s there has been even more warming than during the entire preceding century. Equally important, climate scientists now work with far more reliable models of Earth's climate system than they did 40 years ago (▶ p.68), and they have a better knowledge of the various natural and human factors that influence climate. It is now clear that a natural ice age is not due for many millennia, and we know that the relative impact of aerosols has been small compared to that of greenhouse gas concentrations in recent decades. We also have considerably more data, and we know that the rate of warming in recent decades is greater than can be explained by any natural factors (▶ p.72).

NORTHERN HEMISPHERE CONTINENTAL TEMPERATURE TRENDS

When we compare northern hemisphere temperature trends through the current decade with the shorter record that was available in the mid-1970s (inset), we see that the overall trend is actually toward elevated temperatures, not cooling.

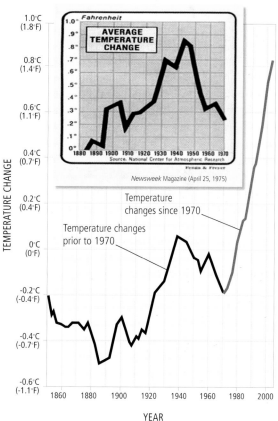

Newsweek Magazine (April 25, 1975)

How does modern warming differ from past warming trends?

Some inaccurate accounts of Earth's climate history make reference to a period called the "Medieval Warm Period." It is sometimes asserted, for example, that because Norse explorers were able to establish settlements in southern Greenland in the late 10th century, global temperatures must have been warmer then than now. Supporters of this view also point to the fact that wine grapes were grown in parts of England in medieval times, indicating that local conditions were warmer than they are today. In fact, the ability of the Norse to maintain colonies in Greenland appears to have been related to factors other than regional climate (such as the maintenance of vigorous trade with mainland Europe), and wine grapes are grown over a more extensive region of England today than they were during medieval times.

Actual scientific evidence

So how do modern temperatures compare to those in past centuries, based on the actual scientific evidence? Evidence from **climate proxy** data

Simulations indicate that the peak warmth during medieval times and the peak cold during later centuries were due to natural factors, such as volcanic eruptions and changes in solar output. By contrast, the recent anomalous warming can only be explained by human influences on climate

(◀ p.42), including tree rings, corals, ice cores, and lake sediments, as well as isolated documentary evidence, indicates that certain regions, such as Europe, experienced a period of relative warmth from the 10th to the 13th centuries, and one of relative cold from the 15th to the 19th centuries (this latter period is often referred to as the "Little Ice Age"). Other regions however, such as the tropical Pacific, appear to have been out of step with these trends.

In fact, the timing of peak warmth and peak cold in past centuries seems to have been highly variable from one region to the next. For this reason, temperature changes in past centuries, when averaged over large regions such as the entire northern hemisphere, appear to have been modest—significantly less than 1.0°C (1.8°F).

Warming everywhere

Unlike the warming of past centuries, modern warming has been globally synchronous, with temperatures increasing across nearly all regions during the most recent century (◀ p.35). When averaged over a large region such as the northern hemisphere (for which there are widespread records), peak warmth during medieval times appears to have reached only mid-20th century levels—levels that have been exceeded by about 0.5°C (about 1°F) in the most recent decades.

Vineyards—in England?
Rows of vines in autumn at Denbies Vineyard, Surrey, England, UK

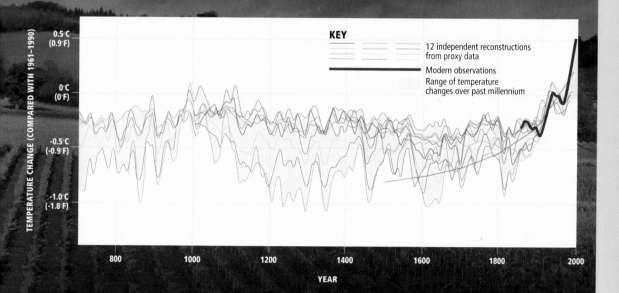

KEY

▭▭▭ ┈┈┈ ╌╌╌ 12 independent reconstructions from proxy data

━━━ Modern observations
Range of temperature changes over past millennium

TEMPERATURE CHANGE (COMPARED WITH 1961–1990)

0.5°C (0.9°F)
0°C (0°F)
-0.5°C (-0.9°F)
-1.0°C (-1.8°F)

800   1000   1200   1400   1600   1800   2000

YEAR

NORTHERN HEMISPHERE TEMPERATURE CHANGES OVER THE PAST MILLENNIUM

A number of independent estimates have been made of temperature changes for the northern hemisphere over the past millennium. While there is some variation within the different estimates, which make use of different data and techniques, they all point to the same conclusion: the most recent warming is without precedent for at least the past millennium

In recent years, scientists have argued for a new term, the **Anthropocene**, to describe the current epoch of Earth history considered to have started with the Industrial Revolution of the past couple of centuries) during which humans have fundamentally changed the planet through modification of the atmosphere, oceans, and Earth's surface. The rise in greenhouse gas concentrations during this epoch is without precedent in many thousands of years (◀ pp.30–31).

Paleoclimate researchers have recently sought to extend the global temperature record further back in time than just the past millennium (◀ pp.48–49), employing longer-term, though coarser,

paleoclimate data such as oce sediments and pollen records

The estimates stretch back th entire current interglacial (rela ice-free) period known as the to the height of the the last ice Last Glacial Maximum, ▶ pp.8 warming since the beginning Industrial Revolution is withou over that entire timeframe.

The warming we will likely see of this century, given business-fossil fuel burning, dwarfs the r temperature variation witnesse entire prior history of human c underscoring the unprecedent modern climate change.

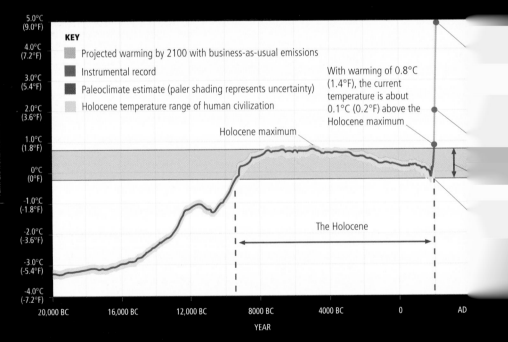

KEY

- Projected warming by 2100 with business-as-usual emissions
- Instrumental record
- Paleoclimate estimate (paler shading represents uncertainty)
- Holocene temperature range of human civilization

With warming of 0.8°C (1.4°F), the current temperature is about 0.1°C (0.2°F) above the Holocene maximum

Holocene maximum

The Holocene

TEMPERATURE ANOMALY

YEAR

LONG-TERM GLOBAL TEMPERATURES

The graph above shows the estimated global temperature change (relative to pre-industria temperatures) from more than 20,000 years ago and projected into the near future, clearly revealing the steep temperature rise since industrialization

City lights

The bright lights of Hong Kong illustrate the huge amounts of energy consumed by modern, industrialized nations. Most of this energy is still produced by burning fossil fuels, which generates greenhouse gases and contributes to global warming.

What can 15 years of western U.S. drought tell us about the future?

Much of western North America, including the western U.S., southwestern Canada, and northwestern Mexico, has been gripped by drought since 1999. Aside from brief respites in late 2004 and 2010, the drought has persisted over the past 15 years. Lack of precipitation, reduced river flows, and lower reservoir levels all confirm the seriousness of the drought.

For much of the region, this is the most persistent and severe drought on record.

This drought is more widespread than the great "Dust Bowl" of the 1930s, which primarily influenced only the central U.S. What is particularly problematic is that the western U.S. has entered into a pattern of severe drought just as demands for its scarce water resources are skyrocketing due to increasing irrigation requirements by agribusiness, and dramatically growing populations in Arizona, Nevada, and Utah.

U.S. Drought Monitor categories
- D4 – Exceptional drought
- D3 – Extreme drought
- D2 – Severe drought
- D1 – Moderate drought
- D0 – Abnormally dry

PATTERN OF U.S. DROUGHT IN AUGUST 2014
Drought conditions existed across much of the western U.S. California experienced its worst drought on record.

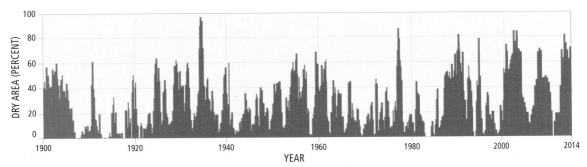

DRY AREA PERCENTAGE IN THE WESTERN U.S.
The most recent interval of drought is unprecedented in modern time.

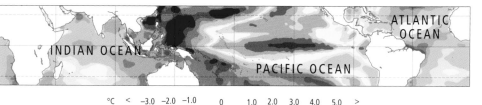

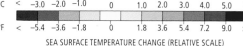

SEA SURFACE TEMPERATURE CHANGE (RELATIVE SCALE)

SEA SURFACE TEMPERATURE (SST) CHANGES IN THE TROPICAL PACIFIC AND INDIAN OCEANS

These SST changes are associated with the enhanced North American drought (based on a calculation for the 1998–2002 period). The concentration of warming in the western tropical Pacific and Indian oceans is reminiscent of the east–west temperature contrast typical of La Niña events.

Ocean temperature effects

Is this drought connected with human-caused climate change? That's a tricky question. Recent research ties the persistent drought conditions in western North America to a pattern of ocean surface temperatures in which the eastern and central tropical Pacific are cool relative to the western tropical Pacific and Indian oceans. This is reminiscent of the "La Niña" pattern. Such a pattern has indeed persisted for most of the past decade and a half, due largely to unusually warm sea surface temperatures in the western tropical Pacific and Indian oceans. The brief cessation of drought during 2004–2005 and again in 2010–2011 was tied to intermittent "El Niño" events, which represent the reverse pattern. In an El Niño event, the eastern and central parts of the tropical Pacific have warm sea surface temperatures, rather than the western parts. Such a pattern favors wetter conditions over much of western North America (▶ p.100).

Lake Mead
Lake Mead is a crucial source of fresh water for the more than 20 million people who live in the desert southwest of the U.S. However, it has been severely depleted due to drought and increased water demand, as can be seen by the lighter "bathtub ring" in this recent photograph.

(Cont.)

Ocean temperature history

So which of the two patterns will climate change favor? Warmer conditions in the western tropical Pacific, and drought? Or more frequent and more severe El Niño events, and wet conditions? Paleoclimate records provide a clue: the data indicate a persistent drought in the western U.S. during the 11th to 14th centuries, and that relatively wetter conditions prevailed during the 15th to 19th centuries. The later wet conditions appear to have been associated with an El Niño–like pattern, while the earlier dry times were associated with an opposite La Niña–like pattern. These changes may have, in turn, been driven by the same natural forces— explosive volcanic eruptions and variations in solar output—that were responsible for the relatively globally warm conditions (◄p.48) of medieval times and the relatively cold conditions of the 15th–19th centuries.

If this association of warmer global temperatures with western North American drought continues in response to human-caused global warming, it could portend increasing troubles for the continent in the future.

EVOLUTION OF GLOBAL DROUGHT PATTERN

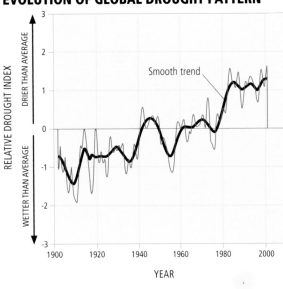

RELATIVE DROUGHT INDEX

DRIER THAN AVERAGE

WETTER THAN AVERAGE

Smooth trend

YEAR

GLOBAL PATTERN OF DROUGHT, AS MEASURED BY THE DROUGHT INDEX

Higher positive values (warm colors) indicate regions suffering more intense drought. The graph above shows how this pattern has evolved over the course of the 20th century. Note the abrupt change in the pattern since the 1980s.

In fact, the drought in the western U.S. is symptomatic of a longer-term, global pattern of increasing drought.

This drought pattern has afflicted much of western North America, the American tropics, most of Africa, Asia, and Indonesia, and parts of Australia. In some cases, the increased drought represents an expansion to higher latitudes of the dry, descending air typically found in the subtropics. In other cases, changes in regional circulation systems such as El Niño (▶ p.100) and the Asian monsoon appear to be important.

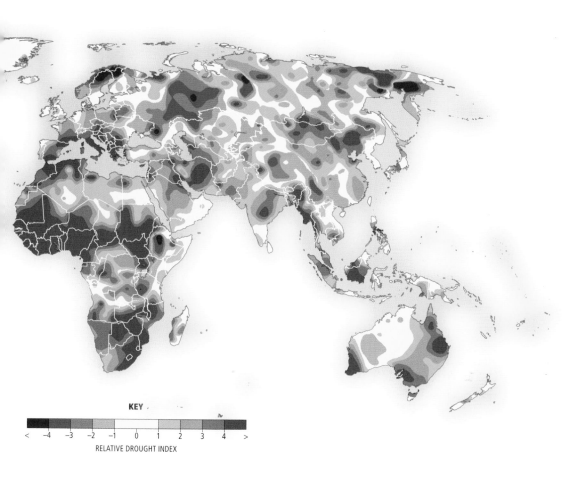

KEY

< −4 −3 −2 −1 0 1 2 3 4 >

RELATIVE DROUGHT INDEX

Signs of things to come?
The 2012 North American heat wave

In June and July 2012, North America experienced an unusually intense and prolonged heat wave. Record-breaking temperatures first gripped the west in late June, spreading eastward and intensifying through early July. Triple-digit temperatures prevailed for weeks over large parts of the great plains, the midwest, and southern U.S. Numerous all-time heat records were broken across the country.

A highly unusual linear band of extreme (130 kmh/80 mph) wind-producing thunderstorms known as a derecho arose from the extreme heat and humidity. As it tracked eastward though the midwest and mid-Atlantic states, it caused hundreds of millions of dollars in damages and left nearly four million people without power and air conditioning as triple-digit temperatures persisted for days. More than 100 people perished as a result of these collective events.

Much like the record heat wave that afflicted Europe in summer 2003, breaking the centuries-old temperature records that had been kept (▶ pp.58–59), there was an important larger context for this event: it was associated with a broad ridge of

high surface pressure, consistent with the predicted expansion of the subtropics to higher latitudes with global warming.

The summer 2012 heat wave took place during what would turn out to be the warmest year on record for the U.S. Moreover, the record heat was part of a persistent pattern. During the prior decade, the rate at which heat records have been broken had risen to more than double what it had been only a half century ago. In other words, it was part of a larger pattern of unusual warmth related to global warming.

The extreme heat had broader impacts and repercussions. Combined with the dry conditions associated with the broad ridge

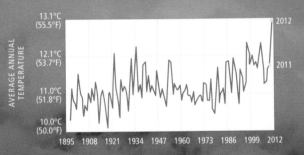

HOTTEST YEAR ON RECORD
In 2012, the contiguous U.S. had its hottest year on record, with every state in the lower 48 experiencing above-average annual temperatures.

Derecho in Kansas
Cars flee the derecho in Kansas in 2012. As the band of severe thunderstorms tracked eastward through the midwest, it resulted in more than 100 deaths and extensive damage to property and infrastructure.

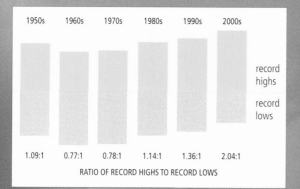

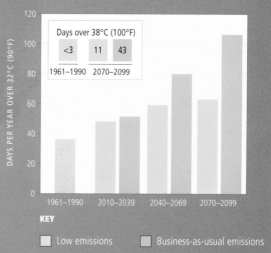

KEY

☐ Low emissions   ☐ Business-as-usual emissions

TEMPERATURE EXTREMES

In the 1950s, the number of record high temperatures was roughly equal to the number of record lows in the contiguous U.S. Since then, the proportion of record highs has increased and there are now about twice as many record highs as record lows.

of high surface pressure, it contributed to drought as widespread and intense as any that has been seen in North America since the dust bowl years of the 1930s.

Impacting 80% of the contiguous U.S., the drought caused widespread crop failure and decimation of agricultural output across America's breadbasket (just the previous year, a similarly disastrous drought had destroyed crops and livestock in Oklahoma and Texas). Fueled by intense heat and drought, wildfires raged across the western U.S., heavily damaging regions such as Boulder and Colorado Springs, Colorado. Extreme drought conditions have persisted through 2014 over a large part of the western U.S. (◀pp.52–53).

TEMPERATURES IN ST. LOUIS, MISSOURI

In St. Louis, Missouri, the center of the 2012 midwestern heat wave, the upward temperature trend is predicted to continue, with potentially more than 100 days a year with temperatures over 32°C (90°F) and more than 40 days a year with temperatures over 38°C (100°F) by the end of the century.

Model simulations show that an event like the 2012 heat wave, considered a record event today, will become typical by the end of the century given business-as-usual burning of fossil fuels. In St. Louis, which was at the center of the 2012 midwestern heat wave, today there are typically fewer than three days a year that exceed 38°C (100°F). Under business-as-usual emissions, that is projected to be more than 40 days (over a month of triple-digit temperatures) by the end of the century.

»

(Cont.)

Like the 2012 North American heat wave, the 2003 European heat wave was associated with an extreme poleward expansion of the high-pressure subtropical zone. Heat waves are predicted to become more common with human-caused climate change (▶ p.112).

In fact, model simulations indicate that this kind of of heat wave would be a roughly once-in-a-thousand-years occurrence. But, when human-induced warming is taken into account, the event is predicted to occur at least once in any given century on average. While this doesn't prove that global warming caused the 2003 European heat wave, it underscores how the increasing occurrence of heat waves in Europe, North America, and elsewhere around the world is consistent with theoretical expectations, and gives us a taste of what is likely to be in store with future climate change.

SUBTROPICAL ZONE EXPANSION

There are two subtropical zones. They lie to the south of the northern polar jet stream, and to the north of the southern polar jet stream. Climate change is predicted to lead to a poleward migration of the polar jet streams, allowing the dry subtropical zones to penetrate farther into mid-latitude regions such as Europe and the U.S. during the summer season.

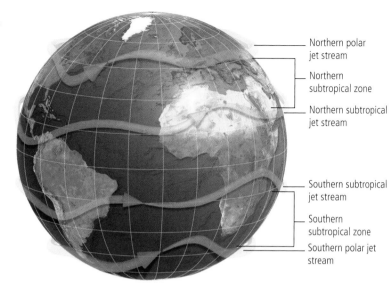

Northern polar jet stream

Northern subtropical zone

Northern subtropical jet stream

Southern subtropical jet stream

Southern subtropical zone

Southern polar jet stream

Model simulations indicate that the 2003 event would have been highly unlikely to occur in the absence of human-induced climate change.

Spreading fires
Firefighters battle one of the many blazes that spread across Portugal during the European heat wave of 2003. Eighteen people were killed in the emergency, which lasted for 17 days over July and August.

Does a cold snap in Peoria invalidate global warming?

The winter of 2013–2014 featured an unusually strong and persistent expansion of the polar front (◀ p.13) and a deep southward dive of the northern hemisphere **jet stream** into the central and eastern U.S. This southward swing brought unusually frigid Arctic air with it, especially during the pronounced cold wave of early January. Yet while conditions were unusually cold over most of the eastern half of the U.S., no all-time cold records were broken. Weather historian Chris Burt of the Weather Channel's *Weather Underground* site

Weather Underground

http://goo.gl/JRWBAS

noted: "The only significant thing about the cold wave is how long it has been since a cold wave of this force has hit for some portions of the country—18 years, to be specific. Prior to 1996, cold waves of this intensity occurred pretty much every 5–10 years." In other words, winter 2013–2014 was what we sometimes refer to as an "old-fashioned winter," at least for part of the U.S.—a part amounting to less than 1% of the global surface area.

The larger context for this event is that most of the rest of North America, and indeed much of the globe, was unusually warm over that same time frame. Alaska

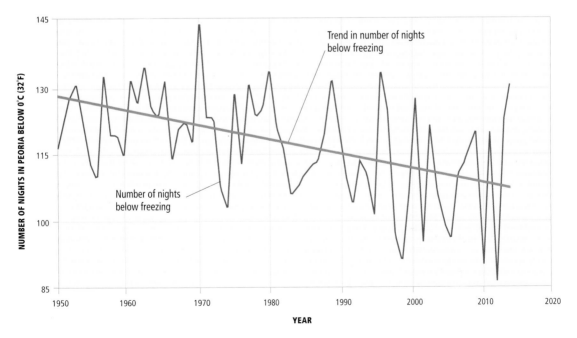

FEWER FRIGID NIGHTS IN PEORIA

Although the winter of 2013–2014 was unusually cold over much of the eastern part of the U.S., the overall trend is for a decrease in extreme cold. This is exemplified by the number of cold nights (under 0°C/32°F) in Peoria, Illinois, which has generally decreased over the past half-century.

LARGE-SCALE TEMPERATURE PATTERN IN JANUARY 2014

This image of surface temperatures on January 23rd, 2014 reveals a large cold area over much of the eastern and northern part of the U.S., with temperatures about 20°C (36°F) below the 1979–2000 average for that date.

-20°C	-15°C	-10°C	-5°C	0°C	5°C	10°C	15°C	20°C
-36°F	-27°F	-18°F	-9°F	0°F	9°F	18°F	27°F	36°F

TEMPERATURE ANOMALY

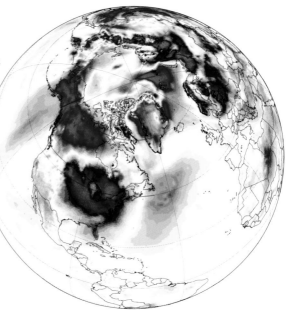

experienced record warmth, and global average temperatures for 2013 had just finished up fourth all-time warmest. And while the boreal winter season December–January was rather cold for the eastern half of the U.S., it was the warmest on record in Australia, and the eighth overall warmest for the globe.

We have already seen that during the past decade the U.S. has been breaking all-time records for warmth at nearly twice the rate we would expect from chance alone (◀ pp.56–57). This is just what models predict to happen as Earth warms up. On the other hand, we certainly wouldn't expect extreme cold to have become more common.

So has it? The answer is a resounding "no." For the vast majority of locations in the U.S., cold extremes have declined significantly as the globe has warmed. Just as we expect, and as climate models predict (▶ pp.112–113), extreme warmth is becoming *more common,* and extreme cold *less common,* as the world continues to warm.

Harsh winter in Peoria
Two men from a snow removal team clear a sidewalk in East Peoria, Illinois, in January 2014, after snow and freezing temperatures swept over the area. During the previous several years, winters in Peoria had been less severe, but the winter of 2013–2014 brought a return to harsher conditions due to the southward swing of the northern hemisphere jet stream.

A tempest in a greenhouse
Have hurricanes become more frequent or intense?

One of the most contentious issues in climate research involves the impacts of climate change on tropical cyclone behavior. A hurricane is defined as an Atlantic (or eastern Pacific) tropical cyclone in which winds sustain speeds greater than 119 kmh (74 mph).

Since modern observations of hurricanes from satellites and aircraft reconnaissance missions are only available for the past few decades, it is difficult to determine whether long-term hurricane trends exist, and whether those trends are associated with human-caused climate change.

One measure of tropical cyclone activity is "accumulated cyclone energy," which takes into account the total energy of storms over their lifetime. Such measurements for the last few decades indicate increases in tropical cyclone activity over several, but not all, major formation basins. In some cases, alternative measures of tropical cyclone activity that quantify the destructive potential or "powerfulness" of storms show dramatic recent increases. These increases appear to be closely related to rising sea surface temperatures in the tropical Atlantic (the only region for which long-term records are available).

There is also evidence of a trend toward greater numbers of tropical cyclones in the tropical Atlantic over the past two decades. A global trend is less clear

2005 Hurricane Season
http://goo.gl/i4mLXd

The eye of the storm
Hurricane Katrina formed on August 23, 2005. It was the costliest hurricane in U.S. history, wreaking havoc along the Gulf Coast from Lousiana to Alabama, and devastating the city of New Orleans. Katrina was also among the deadliest hurricanes in U.S. history, and was one of three Category 5 Atlantic hurricanes that formed during the record-breaking 2005 season.

however, when the numbers are averaged over all of the major hurricane basins.

Basic theoretical considerations as well as detailed climate-model simulations indicate a likely increase in the average intensity of tropical cyclones in all major formation basins (▶ p.115). That said, it is uncertain how the actual number of tropical cyclones is likely to be affected by climate change, and this remains an area of active climate research.

Hurricane Sandy made landfall on the New Jersey coast as a powerful storm in late October 2012. It set a number of records, including the lowest

Superstorm Sandy
http://goo.gl/MY7ZuG

central pressure (a measure of strength) for a storm north of Cape Hatteras, NC. It was also the largest hurricane to form in the Atlantic, and led to a record 4 m (13 ft) storm surge at Battery Park in New York City. The storm did an estimated $65 billion worth of damage due to extensive flooding and wind-caused destruction along the New Jersey coast and in and around the New York City area. While global warming did not cause the storm, it arguably made it worse: record sea surface temperatures (more than 5°C/9°F above average) along the U.S. east coast contributed to the unusual strength and flooding potential of the storm, and global sea level rise (▶ pp.110–111) added roughly 30 cm (1 ft) to the coastal surge, leading to an estimated extra 65 square kilometres (25 square miles) of flooded area.

SEA SURFACE TEMPERATURE VS POWERFULNESS IN TROPICAL ATLANTIC CYCLONES

Sea surface temperatures are closely related to the powerfulness of tropical cyclones. The dramatic increase in cyclone powerfulness in recent decades closely parallels the anomalous rise in sea surface temperature in the tropical Atlantic region.

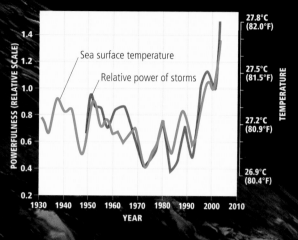

ACCUMULATED CYCLONE ENERGY IN TROPICAL CYCLONE-PRODUCING BASINS

There is a clear upward trend in cyclone energy for the tropical Atlantic in recent years. Trends are less clear for cyclones in the Pacific and Indian oceans.

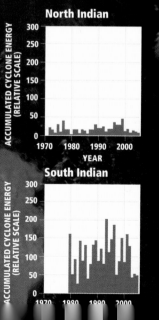

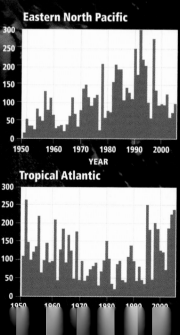

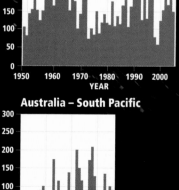

The vanishing snows of Kilimanjaro
An icon of climate change?

There is perhaps no snowcapped peak in the world as iconic as Mount Kilimanjaro, Tanzania, immortalized in Ernest Hemingway's *The Snows of Kilimanjaro*. A remarkable group of perennial ice fields can be found at altitudes of roughly 5 km (3 miles) and above atop Mount Kilimanjaro, at a latitude only a few degrees south of the equator.

Like mountain glaciers in many parts of the world, the snows of Kilimanjaro provide a key source of fresh water. The ultimate demise of Kilimanjaro's ice, projected to take place sometime within the next two decades at current rates of decline, consequently poses a threat to the people who inhabit the region.

Are the vanishing snows of Kilimanjaro a casualty of global warming?

We know that mountain glaciers the world over are disappearing. This disappearance is generally related to increased melting due to warmer atmospheric temperatures. In fact, at the altitudes where most tropical mountain glaciers are found, the atmosphere has warmed even more in recent decades than Earth's surface.

1912
Average snow area: 12.0 square km/4.6 square miles

1970
Average snow area: 5.0 square km/1.9 square miles

2000
Average snow area: 2.5 square km/0.9 square miles

2014
Average snow area: 1.5 square km/0.6 square miles

Kilimanjaro has had permanent ice fields for 12,000 years: it is unlikely that their current dramatic retreat is a coincidence.

However, it is not as simple as saying warmer temperatures cause more ice to melt. While some melting has been observed recently on Kilimanjaro, ice loss at these latitudes and elevations occurs mostly through "sublimation" (the conversion of ice directly into the atmosphere), rather than by melting.

The rate of sublimation is influenced by factors in addition to temperature, such as cloud cover and humidity. Also, the amount of ice that exists on mountain glaciers isn't controlled by melting or sublimation alone, but represents a balance between the rate of ice loss due to those processes, and the rate of accumulation of ice. Accumulation is determined by the amount of precipitation that falls each year, so it is likely that decreased snowfall in the region has had a significant impact on the extent of the ice fields. The changing precipitation patterns, leading to drier conditions in the region, are tied to larger-scale climate changes. In this sense, human influence on climate is probably responsible for the imminent demise of Kilimanjaro's snows, even if warmer regional temperatures alone are not.

Mount Kilimanjaro
Kilimanjaro, in Tanzania, Africa, is one of a number of locations around the world where ancient mountain glaciers are disappearing before our eyes.

ICE ON KILIMANJARO SINCE 1912

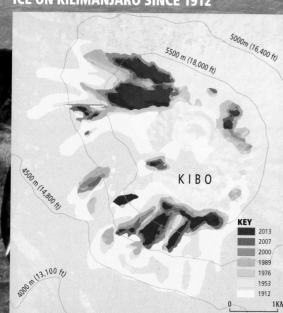

5000m (16,400 ft)

5500 m (18,000 ft)

KIBO

4500 m (14,800 ft)

4000 m (13,100 ft)

KEY
2013
2007
2000
1989
1976
1953
1912

0   1KM

The last interglacial
A glimpse of the future?

Driving south on U.S. Route 1 from Miami, Florida, you pass a road sign for the Windley Key Fossil Reef Geological State Park, the site of a former limestone quarry. Here, you are at the highest point in the Florida Keys—islands that 125,000 years ago were an impressive chain of coral reefs. At that time, sea level was about 6 m (20 ft) above where it is today, most likely because the Greenland and West Antarctic ice sheets were smaller and much of the water they contain was at that time in the ocean, not locked up in glacial ice as it is today. The climate then began to cool, and the reefs became exposed as the sea level fell when water from the ocean evaporated and froze to form ice sheets in the northern hemisphere. At the height of the last glaciation, the seas had retreated 10 km (6 miles), and the Keys stood 120 m (400 ft) above the ancient sea level. Now they are barely a few meters (several feet) above sea level, with Windley Key at the highest elevation (6 m/20 ft). These fluctuations in sea level are the result of the 40-thousand- to 100-thousand-year "glacial–interglacial" cycles of the last 2 million years. These cycles occur in conjunction with the repeated growth and demise of the almost 2-km/1.24-miles-thick North American ice sheets that covered most of Canada and the northern United States, as well as other smaller ice sheets in northern Europe.

Climate in the last interglacial interval

Indeed, with modern global warming causing sea levels to rise again, low-lying areas like the Florida Keys are gradually disappearing (▶ p.110). To understand the future implications of these changes, scientists naturally turn to time periods with similar conditions—like the last interglacial—for answers to questions such as how much the Greenland ice sheet is likely to shrink.

A wealth of information on the climate of the last interglacial has been collected in recent decades. Ice cores (▶ p.88) reveal that atmospheric CO_2, methane, and nitrous oxide levels were close to pre-industrial levels. Coastal ocean temperatures were

Will the Florida Keys turn back to coral reefs once again?

generally warmer than today, there was considerably less sea ice surrounding places like Alaska, and boreal forests had overtaken regions that are now tundra in Siberia and Alaska. Arctic summers were warmer than pre-industrial summers by 5°C (9°F).

Earth's orbit affects temperature

Why was the last interglacial so warm— warm enough to melt the Greenland ice sheet? We know from ice-core data that it wasn't due to higher CO_2 levels. The answer is that the northern hemisphere received 10% more solar radiation than it does today, not because the Sun was brighter, but because Earth's orbit around the Sun was different than it is today. Earth's orbit changes slowly and regularly in response to the tug of the Sun, Moon, and the large planets, especially Jupiter. The orbital configuration 125,000 years ago provided more direct rays to northern latitudes in the summer because the tilt of Earth's spin axis was greater.

Even though this wasn't a greenhouse warming event, and thus not a direct analogy, the information we can learn from the last interglacial suggests that the Greenland ice sheet is subject to considerable shrinkage from relatively subtle changes.

Florida Keys
U.S. Route 1 extends over a fossil coral reef that formed 125,000 years ago.

EARTH'S CHANGING ORBIT

Earth's orbit and rotation are not constant, but change cyclically over many millennia. Its orbit varies from elliptical to circular, and the planet also tilts about its axis and wobbles as it rotates. Over time, these changes affect temperature, and they can be correlated to climate swings and glacial– interglacial cycles.

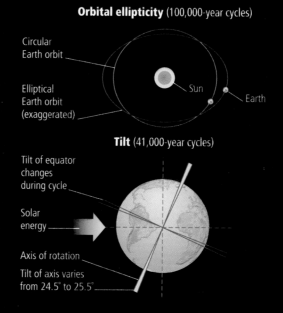

Orbital ellipticity (100,000-year cycles)

Circular Earth orbit

Elliptical Earth orbit (exaggerated)

Sun

Earth

Tilt (41,000-year cycles)

Tilt of equator changes during cycle

Solar energy

Axis of rotation

Tilt of axis varies from 24.5° to 25.5°

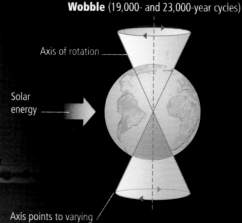

Wobble (19,000- and 23,000-year cycles)

Axis of rotation

Solar energy

Axis points to varying positions in space

Earth-Sun Relations

http://goo.gl/KjZ8BH

How to build a climate model

Building a model of Earth's climate is a challenging endeavor because the climate is governed by many complex physical, chemical, and biological processes and their interactions. Earth's climate can be thought of as a system with different physical properties (the oceans, atmosphere, and the ice sheets and glaciers in polar and high-elevation regions). Each of these components, and the interactions between them, are governed by the laws of physics: fluid dynamics, thermodynamics, and radiation balance. There are many further complications, however, that must be taken into account in modeling Earth's climate. For example, life on Earth (the **biosphere**), which includes both plants and animals, plays a key role. The biosphere is involved in the global recirculation of water and carbon, and it influences the composition of the atmosphere and Earth's surface properties—all of which impact climate.

Simple climate models

The simplest models ignore the three-dimensional structure of Earth, atmosphere, and oceans, and simply focus on the balance between incoming solar energy and outgoing terrestrial (heat) energy. It is the balance between these incoming and outgoing sources of energy that determines temperatures on Earth. Even in these simple models, the greenhouse effect (◄ p.22) must be accounted for. This is usually accomplished through a modification that represents the way heat is absorbed and emitted by the atmosphere. It is also essential, even in simple models, to account for feedback loops (◄ p.24) that can either amplify (positive feedback) or diminish (negative feedback) the impacts of any changes. In most climate models, the net impact of feedbacks roughly doubles the magnitude of the expected warming or cooling response to imposed changes.

BUDGETING THE INCOMING RADIATION

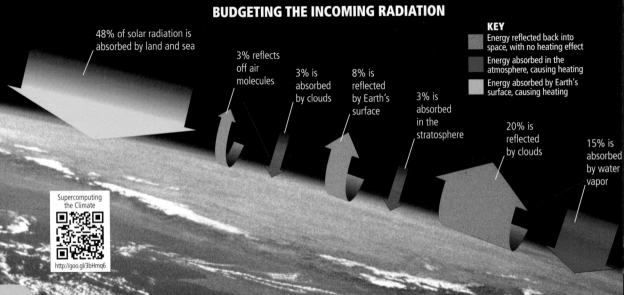

48% of solar radiation is absorbed by land and sea

3% reflects off air molecules

3% is absorbed by clouds

8% is reflected by Earth's surface

3% is absorbed in the stratosphere

20% is reflected by clouds

15% is absorbed by water vapor

KEY

Energy reflected back into space, with no heating effect

Energy absorbed in the atmosphere, causing heating

Energy absorbed by Earth's surface, causing heating

Supercomputing the Climate

http://goo.gl/3bHmq6

Complex climate models

The most complex climate models, referred to as **general circulation models (GCM)**, take into account the full three-dimensional structure of the atmosphere and oceans, arrangement of the continents, details of coastlines and ocean basins, and surface topography. These models calculate not only surface temperatures, but also other important climate variables, such as precipitation, atmospheric pressure, surface and upper level winds, ocean currents, temperatures, and salinity. This is accomplished by breaking the oceans and atmosphere into many small grid boxes, and by using the underlying physical, chemical, and biological relationships to calculate values for properties of each box and the interactions between different boxes.

Should climate model predictions be trusted?

Current climate models do a remarkably good job of reproducing key features of the actual climate such as the jet streams in the atmosphere, the seasonal band of rainfall and cloudiness that migrates north and south of the equator, and even the complex internal climate oscillation associated with the El Niño phenomenon (▶ p.100). These models also closely reproduce

past changes (▶ p.72), including the demise of Arctic sea-ice extent, intermittent cooling episodes caused by volcanic eruptions, and past glacial and hothouse climate states. Large-scale rainfall patterns are reasonably well simulated, but finer-scale details are a continuing challenge for the models. Nevertheless, significant progress is being made, and we have good reason to take their projections of possible future changes in climate seriously.

Atmosphere is divided into 3-D grid boxes, each with its own local climate

Air in grid boxes interacts horizontally and vertically with other boxes

Influence of vegetation and terrain is included

Water in oceanic grid boxes interacts horizontally and vertically with other boxes

Oceanic grid boxes model currents, temperature, and salinity

LOST ENERGY

Only 48% of incoming solar energy reaches Earth's surface to heat the continents and oceans. Nearly one-third of the total energy that encounters the atmosphere is immediately returned to space—reflected by clouds or air molecules. Additional radiation is reflected by Earth's surface, especially in icy regions. The remainder is absorbed by stratospheric gases

COMPLEX CLIMATE MODELLING

Global climate models divide a region like Spain into a number of grid cells and calculate the energy, moisture, and carbon budgets for each cell. Cells are influenced by the surrounding cells and by incoming solar energy, outgoing Earth energy, and water and carbon fluxes between ocean or land surface and atmosphere. Properties like temperature, humidity, and cloud cover are treated as uniform for the whole grid cell. Thus, higher-resolution models with smaller grid cells tend to perform better but require considerably greater

Profiles in climate change science

James Hansen

James Hansen is a well-known climate scientist and former Director of NASA's Goddard Institute for Space Studies (GISS), a major climate-modeling center. Hansen, a member of the U.S. National Academy of Sciences, came to prominence during his congressional testimony in the hot summer of 1988, when he became the first scientist to publicly testify before the U.S. Congress that the effects of human-caused climate change were apparent. His testimony seems prescient when we look at what has happened since his now-famous pronouncement.

James Hansen Profile

http://goo.gl/gSd8sg

Find out more

For more information about Hansen's work, and to see how his global warming predictions from the late 1980s have panned out, scan the QR code.

Susan Solomon

Susan Solomon is a leading atmospheric scientist and a recipient of the highest honor granted to scientists in the U.S., the Presidential Medal of Science. She is also a member of the National Academy of Sciences. Solomon's early scientific contributions involved determining the subtle atmospheric chemistry underlying Antarctic stratospheric ozone depletion. She recognized the role of polar stratospheric clouds (PSCs) in ozone destruction by discovering that they provide a surface that allows chemical reactions to proceed more rapidly than had been previously known.

Susan Solomon Profile

http://goo.gl/PjsLXX

Find out more

To read more about Solomon's recent contributions to the study of atmospheric chemistry and climate, scan the QR code.

Stephen Schneider

Steven Schneider (1945–2010) was both a leading climate scientist and a leading public communicator of the science, impacts, and risks of human-caused global warming. He was one of the first climate scientists to attempt to quantify the competing roles of human greenhouse gases and industrial particulates on global temperatures. He was also one of the scientists to speak up about the threat of human-caused climate change. Schneider was a recipient of the Award for Public Understanding of Science and Technology from the American Association for the Advancement of Science, a MacArthur Award, and was elected to the U.S. National Academy of Sciences.

Stephen Schneider Profile

http://goo.gl/Gk4xo9

Find out more

To learn more about Schneider's important contributions to our understanding of climate change, scan the QR code.

Warren Washington

Warren Washington, one of the world's leading climate scientists, pioneered the development in the 1960s of early generation climate models (GCMs) designed to represent the three-dimensional nature of atmospheric circulation and energy balance. He continues to publish leading scientific articles using state-of-the-art simulations to assess future climate change. Washington was awarded the Dr. Charles E. Anderson Award of the American Meteorological Society; he was elected to the National Academy of Engineering; and he received the Presidential Medal of Science from President Barack Obama.

Warren Washington Profile

http://goo.gl/eyVvwr

Find out more

For more about Washington's work advising policy makers and as a mentor for individuals, educational programs, and outreach initiatives, scan the QR code.

Comparing climate model predictions with observations

The average annual temperature of the planet is not expected to be constant in the absence of human influence. Variations in solar energy input warm and cool the planet on yearly, decadal, and longer timescales, while volcanic eruptions cool the planet from months up to years after a major eruption.

Using climate models

We can use climate models to calculate how natural variations in solar energy and volcanic aerosols alone would have driven climate change over the last century if there hadn't been any human influence and compare the result to the observed record (graph 1). Then we can

PREDICTED/OBSERVED CLIMATE TRENDS

1 **Predicted temperature trends from models (average shown with red and blue lines and ranges shown with gray and yellow shading), taking into account the impacts of natural forces alone, compared to observation (black line).**

Thirteen different climate models indicate which portion of the annual average temperature variations over the last century can be attributed to natural forces alone. Natural forcing does not explain the warming since the 1970s.

2 **Predicted temperature trends from models taking into account the impact of rising greenhouse gas concentrations alone (average model result shown with red line and ranges in yellow), compared to observation (black line).**

The expected warming from greenhouse gases alone is greater than that observed.

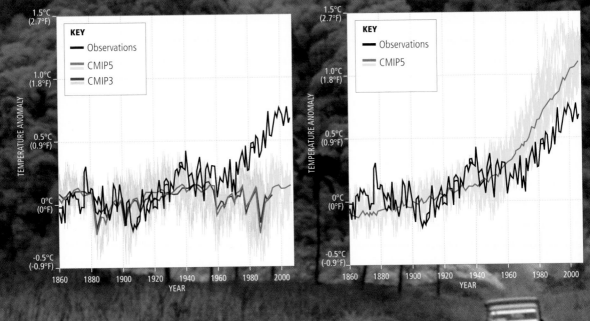

Graph 1 — KEY: Observations, CMIP5, CMIP3. Axis: TEMPERATURE ANOMALY; 1.5°C (2.7°F), 1.0°C (1.8°F), 0.5°C (0.9°F), 0°C (0°F), -0.5°C (-0.9°F). YEAR: 1860, 1880, 1900, 1920, 1940, 1960, 1980, 2000.

Graph 2 — KEY: Observations, CMIP5. Axis: TEMPERATURE ANOMALY; 1.5°C (2.7°F), 1.0°C (1.8°F), 0.5°C (0.9°F), 0°C (0°F), -0.5°C (-0.9°F). YEAR: 1860, 1880, 1900, 1920, 1940, 1960, 1980, 2000.

include the effect of human-produced greenhouse gases, and see if the model predictions fit the observed data better (graph 2). Finally, we can put together both the natural and human forces, and see how well the model fits the observed temperatures (graph 3). The temperature deviations in the graphs below are expressed relative to the average temperatures from 1901 to 1997. The good fit between actual observations and models that take into account human actions and natural forces gives us confidence that we can predict future climate responses to fossil-fuel burning.

3 **Predicted temperature trends from models (line color and shading as in graphs 1 and 2) that take into account the impacts of both natural and anthropogenic factors, including the cooling effect of anthropogenic aerosols.**

When human forces, especially the buildup of atmospheric greenhouse gases and industrial aerosols, and natural forces are applied simultaneously to these same climate models, the general trends and many of the anomalies are reproduced.

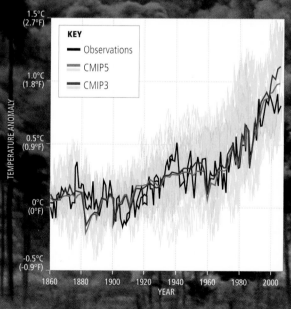

Mount Pinatubo explodes
Pinatubo began erupting in the Philippines on June 12, 1991. It caused a subsequent 0.5°C (0.9°F) cooling of the atmosphere.

Regional vs global trends

It is now possible to tie temperature changes over the oceans, bodies of land, and at the regional scale of individual continents to human activity. When scientists compare observed and modeled temperature changes at these regional scales, they find that the warming trends observed for individual continents, such as North America, Europe, and Africa, cannot be explained by natural factors like volcanoes and changes in solar output. As we have seen on a global scale (◀ p.72), only when the models include the human, or "anthropogenic," component—warming due to increased greenhouse gas concentrations and the more minor cooling impact of industrial aerosols—can they explain observed regional warming trends (see graphs at right). This indicates that human influences are now having a detectable impact on temperature changes measured in individual regions.

Temperature changes, however, are just one of many significant regional impacts. Of equal or greater significance for human society are projected changes in drought and rainfall patterns.

Observing Earth
This composite "blue marble" view of Earth was taken in 2012 using the VIIRS (Visible/Infrared Imager Radiometer Suite) on the Suomi NPP satellite. This satellite carries a range of instruments that enable it to measure a variety of weather and climate variables, such as sea, land, and atmospheric temperatures, ozone levels, and radiation reflection and emission.

GLOBAL TRENDS. LAND VS OCEAN

These graphs compare actual observations and model results both with and without human factors included. Only the models that take into account both human and natural factors make predictions that look like actual data trends.

Global land and ocean surface

KEY

Full spread of models, taking into account natural factors only

Average of models, taking into account natural factors only

Full spread of models, taking into account both human and natural factors

Average of models, taking into account both human and natural factors

Observations

Observations where data coverage is incomplete

REGIONAL CONTINENTAL TRENDS

These graphs compare actual observations and model results both with and without human factors included. As on the global scale, only models taking into account both human and natural factors make predictions that correspond to actual data.

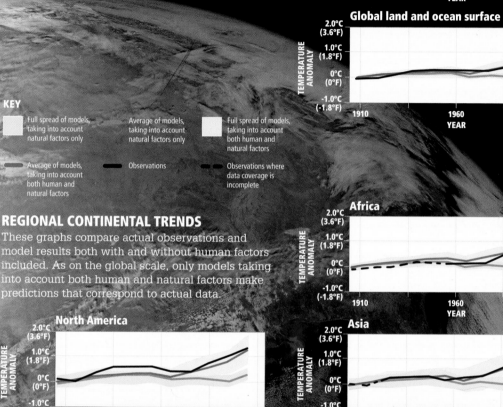

North America

Africa

Asia

South America

Australia

Europe

Antarctica

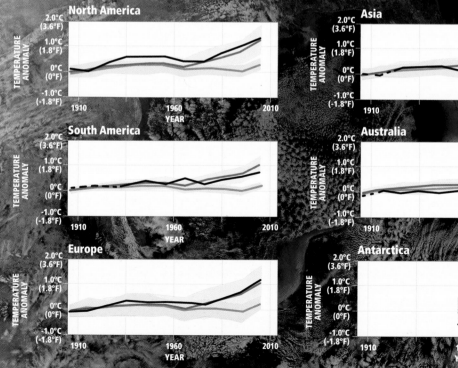

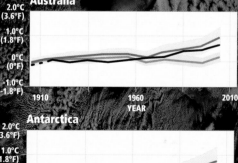

Some climates disappear as others emerge...

There are different ways to classify Earth's land regions on the basis of climate. The most widely used is the Koeppen climate classification. Climate zones are defined by their prevailing seasonal temperature and precipitation patterns, and a shorthand is used for each zone. For example, tropical climates are denoted by "A" with a suffix that denotes whether they are rainforest (Af), monsoonal (Am), or wet-and-dry (Aw). Temperate climates are denoted by "C," while regions with "continental" climates characterized by warm summers and cold winters are denoted by "D." Dry climates are denoted by "B."

The contiguous U.S. is characterized by a number of distinct climates: the humid subtropical temperature climate (Cfa) in the southeast, continental climates (Dfa and Dfb) in the northeast and north central states, and dry climates in the interior mountain and western states. The very southern tip of Florida is characterized as a tropical wet-and-dry (Aw) climate, while the west coast is characterized by a warm-summer dry-summer subtropical "Mediterranean" climate (Csa) in the south, and a temperate "Marine West Coast" climate (Cfb) in the north.

Most of Europe is characterized as humid subtropical (Cfa), but regions neighboring the Mediterranean are classified as "Mediterranean" climates (Csa and Csb). Western Europe is characterized by a marine west coast climate (Cfb); the eastern parts of Europe are "continental" climates (mostly Dfb); and the most northern (and extreme alpine) regions are classified as "polar climates" (ET and EF). These climates exist in other regions as well (see map next page).

As climate change shifts seasonal temperature and precipitation patterns around the world, "novel" climates that do not currently exist in classification schemes

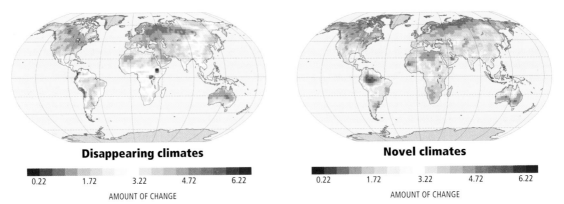

Disappearing climates

0.22   1.72   3.22   4.72   6.22

AMOUNT OF CHANGE

Novel climates

0.22   1.72   3.22   4.72   6.22

AMOUNT OF CHANGE

DISAPPEARING AND NOVEL CLIMATES

These two maps show most of the present climates expected to disappear (left) and novel climates expected to appear (right) by the end of the 21st century given businesss-as-usual carbon emissions. The scale is relative, with numbers over 3.22 indicating likely change.

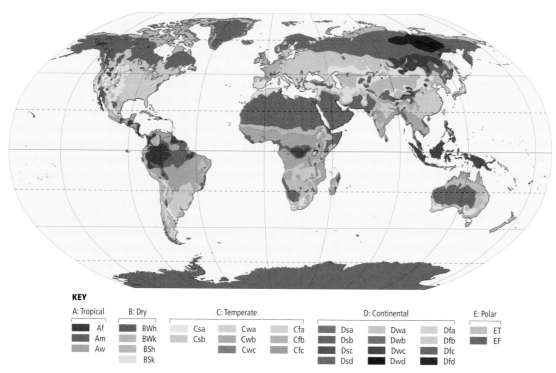

KEY

A: Tropical		B: Dry		C: Temperate					D: Continental								E: Polar	
	Af		BWh		Csa		Cwa		Cfa		Dsa		Dwa		Dfa			ET
	Am		BWk		Csb		Cwb		Cfb		Dsb		Dwb		Dfb			EF
	Aw		BSh				Cwc		Cfc		Dsc		Dwc		Dfc			
			BSk								Dsd		Dwd		Dfd			

KOEPPEN CLIMATE ZONES

The Koeppen classification categorizes climates into five main zones (denoted by the initial letters A through E), and then into a number of smaller zones (denoted by the subsequent letters), as outlined in the main text (left).

are predicted to emerge, while some that exist today will likely disappear. The risk to novel and disappearing climates is unevenly distributed across the world's climate zones. Novel climates are particularly likely in tropical regions where high temperatures may exceed the range found today. Climates most likely to disappear are in alpine or boreal environments where the coldest temperatures that prevail today will no longer exist.

The preferential loss of climate from certain zones could have dire consequences for biological diversity (▶ pp.130–131). The loss of the Arctic sea ice habitat poses a threat to the polar bear, while the disappearance of alpine tundra environments in North America threatens the American pika.

Tropical species may especially be threatened as tropical temperatures vary over a smaller range and the natural ranges of animal species are inherently smaller nearer the equator. Small shifts in temperature may thus pose a greater adaptive challenge. The Amazon basin is at particular risk due to the added challenges of increased wildfire occurrence and forest loss due to the predicted aridification of the region.

Scientists are particularly concerned because many of the areas where existing climates may disappear (the Andes, Mesoamerica, southern and eastern Africa, the Himalayas, and the Philippines) are regions of great current biological diversity. The potential for species loss is thus particularly great in these regions.

"Fingerprints" distinguish human and natural impacts on climate

Not all factors that affect climate have the same pattern of influence on temperature. One way to distinguish natural and human sources of modern climate change is to look for particular "fingerprints"—that is, specific spatial patterns of change that help to identify the likely underlying factor associated with that change.

Natural impacts

Two natural factors have influenced climate change over the past century:

- Changes in solar output

- Explosive volcanic activity

Human impacts

Major human impacts on climate include the greenhouse gas emissions resulting from:

- Fossil-fuel burning (primary impact)

- Industrial aerosols (secondary impact) (◀ p.19)

We have seen that on both the global (◀ p.72) and regional scale (◀ p.74) climate model simulations that include all of the natural and human factors together can do a fairly good job of reproducing the observed pattern of surface warming. By contrast, simulations that include only natural factors are unable to reproduce the spatial pattern of observed surface warming. This observation holds when we compare simulations at the detailed spatial level, as seen in the global maps

Simulations that include only natural factors are unable to reproduce the spatial pattern of the observed surface warming.

Coal fire
Coal-fired power stations are among the primary sources of industrial greenhouse gas emissions.

Model-predicted trend in annual average surface temperature 1979–2005, including natural factors only

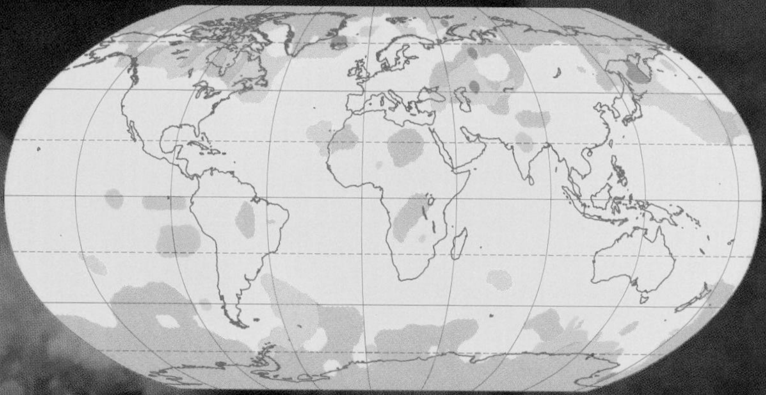

Model-predicted trend in annual average surface temperature 1979–2005, including natural and human factors

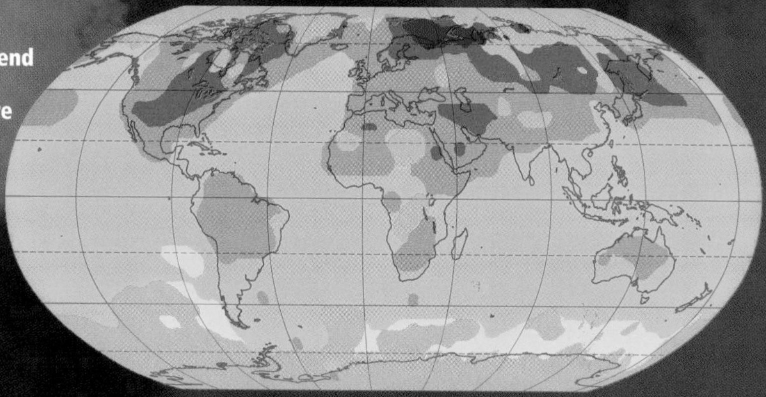

Actual trend in annual average surface temperature 1979–2005

WARMING PATTERNS

The pattern of warming over the past few decades is not reflected in the models that only take into account the impacts of natural forces alone (top map). Instead, actual observations (bottom map) correspond closely with model predictions that take into account the impacts of both natural and human forces (middle map).

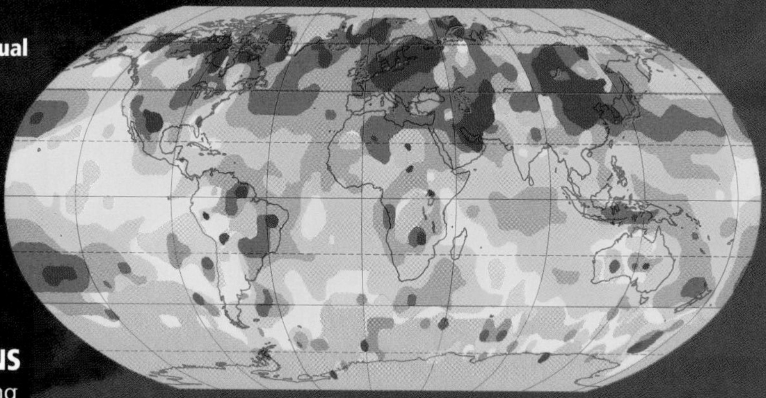

Surface temperature key

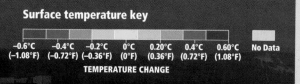

| −0.6°C (−1.08°F) | −0.4°C (−0.72°F) | −0.2°C (−0.36°F) | 0°C (0°F) | 0.20°C (0.36°F) | 0.4°C (0.72°F) | 0.60°C (1.08°F) | No Data |

TEMPERATURE CHANGE

(Cont.)

other fingerprint is the pattern of expected atmospheric temperature changes. Different factors, natural and man-generated, have different effects on the different layers of the atmosphere (p.38). Increases in solar output are predicted to warm essentially the entire atmosphere from top to bottom. Volcanoes cool the lower atmosphere (the troposphere) slightly, and warm the mid-level atmosphere (the stratosphere).

By contrast, human-generated increases in greenhouse gases are predicted to warm the lower atmosphere (the troposphere) substantially, at the expense of cooling the upper atmosphere (the stratosphere and above).

Human-generated greenhouse gas concentration increases are thought to be the primary cause of atmospheric temperature changes over the past century. Not surprisingly then, the

Solar effect on atmospheric temperature

Volcanic effect on atmospheric temperature

Human-generated greenhouse gas effect on atmospheric temperature

ATMOSPHERIC TEMPERATURE CHANGE

The model-predicted pattern of atmospheric temperature change taking into account the impacts of both natural and human influences (far right) looks much like the greenhouse gas pattern alone (second from right). This is because human-generated increases in greenhouse gas concentrations are believed to have dominated atmospheric temperature changes over the past century (shown are the model-simulated trends in annual average

Atmospheric temperature key

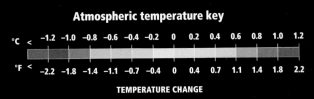

| °C | < | −1.2 | −1.0 | −0.8 | −0.6 | −0.4 | −0.2 | 0 | 0.2 | 0.4 | 0.6 | 0.8 | 1.0 | 1.2 |
| °F | < | −2.2 | −1.8 | −1.4 | −1.1 | −0.7 | −0.4 | 0 | 0.4 | 0.7 | 1.1 | 1.4 | 1.8 | 2.2 |

TEMPERATURE CHANGE

Atmospheric layers are not drawn to scale; height has been exaggerated in order to show color variations as clearly as possible.

expected pattern of atmospheric
temperature change due to all impacts
combined (natural and human) looks much
like the greenhouse gas pattern alone.
Indeed, we see that this pattern matches
the observed pattern of temperature
change in recent decades (◀ p.38).

**Combined effect of
human and natural
forces on atmospheric
temperature**

Natural fireworks
Lava and plumes of smoke shoot
into the air during an erruption
north of Vatnajokull, Iceland's
most extensive glacier.

Part 2
Climate Change Projections

Projections of how Earth's climate will change are uncertain. They depend on both the unknown future trajectory of greenhouse emissions and the uncertain response of the climate to these emissions. Nonetheless, researchers can draw certain conclusions given best-guess scenarios of fossil-fuel burning and the average projections of theoretical climate models. Scientists can project, for example, that for "middle-of-the-road" emissions scenarios, the globe is likely to warm by several more degrees by the end of the 21st century. This warming is likely to be associated with a dramatic decrease in Arctic sea ice, an acceleration of sea level rise, and increased drought, flooding, and extreme weather for many regions of Earth.

Projected changes to climate

• Forecasts of future CO_2 emissions lead to a range of possible scenarios. The middle-of-the-road projection will see the atmosphere warm by 1.0–2.6°C (1.8–4.7°F) between 2000 and 2100.

• The pattern of warming will likely be uneven, with the greatest warming in the polar areas of the northern hemisphere and more warming over land than the oceans.

• Projections suggest that, in general, areas that are already moist will become more moist, while dry areas will become drier.

• The precise pattern of change will depend on natural irregular oscillations of the climate system. Some changes may be sudden and irreversible, as the climate system passes tipping points.

Nature's response to CO_2

• The planet's response will partly be determined by positive and negative feedback loops. Positive feedback loops are expected to outweigh negative ones, leading to more rapid CO_2 buildup and greater warming.

• Sea level is expected to rise as continental ice sheets melt. The latest projections suggest a rise of 0.5–1.2m (2–4ft) by 2100.

• As climate change accelerates, we can expect fewer frosty days, longer heatwaves, and more intense rainstorms.

How sensitive is the climate?
Modern evidence

To determine the potential magnitude of future global warming, scientists find it useful to estimate something they call **climate sensitivity**.

▪ **Climate sensitivity** defines the amount of warming (in degrees Celsius) that we can expect to occur when there is a change in the factors that control climate. This definition places a numerical value on how much our planet will warm in response to future increases in greenhouse gas emissions. Climate sensitivity is typically expressed in terms of the expected surface warming that will occur in response to a doubling of atmospheric CO_2 levels from their pre-industrial level of roughly 280 parts per million (ppm) by volume.

▪ **Equilibrium climate sensitivity** takes into account the fact that the full amount of warming in response to an increase in greenhouse gas concentrations may not be realized for many decades, due to sluggish ocean warming. In plain language, this means that if we say equilibrium climate sensitivity is 3.0°C (5.4°F), we mean that Earth will eventually warm by 3.0°C (5.4°F) if CO_2 levels reach 560 ppm by volume. At current rates of fossil-fuel burning, this doubling of CO_2 levels is expected to occur midway through the 21st century. However, the resulting warming may not fully be experienced until at least 2100.

So how do scientists estimate climate sensitivity?

Using climate models, scientists compare observed temperature changes (from the **instrumental record** of the past 160 years) with simulations of temperature changes over this same time frame (◀ p.72). In order to determine the actual climate sensitivity, certain types of climate models are tuned to different climate sensitivity values. Scientists then determine which of these sensitivity values best match the observed temperature changes. By looking at the various climate sensitivities that fit reasonably well with the actual temperature record, the scientists can quantify the uncertainty of the estimated climate sensitivity values. This range is quite large if only the relatively short instrumental record is used.

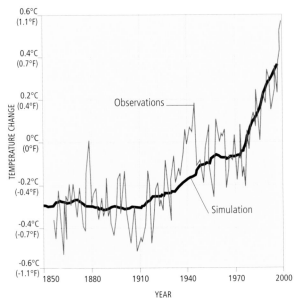

SIMULATED VS ACTUAL SURFACE TEMPERATURE CHANGES

A climate sensitivity of 2.7°C (4.9°F) was assumed in these simulations, which took into account both human and natural factors over a 150-year-long interval during which the simulations overlap with observational temperature records (shown are the model-simulated trends in annual average temperatures from 1850 to 1999).

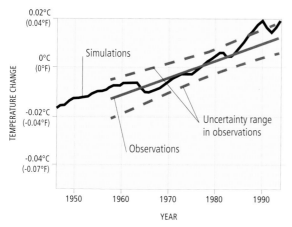

SIMULATED VS ACTUAL DEEP-OCEAN TEMPERATURE CHANGES

A climate sensitivity of 2.9°C (5.2°F) was assumed in these simulations, which took into account both human and natural factors during a 50-year interval for which deep-ocean (below 1000 m/3280 ft depth) temperature observations exist.

Floating monitors
A global array of floats, such as the Argo one shown here, gather data to keep scientists informed of climate changes on the surface of the oceans. They measure variables such as air and ocean surface temperature, sea level pressure, wind speed, and wave heights.

Deep ocean temperature

Over the past few decades, it has also been possible to make use of temperature measurements from the deep oceans. While such data are useful, they are limited by an even shorter available record.

Do we need more data?

Overall, the various observations suggest an equilibrium climate sensitivity in the range of 3.0°C (5.4°F). In other words, it is estimated that Earth's surface will warm by 3.0°C (5.4°F) if CO_2 concentrations double. Uncertainties, however, are large.

Modern instrumental observations could be consistent with a climate sensitivity anywhere from 1.5°C to 9.0°C (2.7°F to 16.2°F).

With such a wide temperature range, the effect of climate change could be anything from essentially negligible to wholly catastrophic. This uncertainty is inevitable when we only have access to a short (roughly century-and-a-half) record. There are so many different natural and human factors that are simultaneously at play, and each has impacts that are individually uncertain. For this reason, scientists turn to other longer-term sources of information. In the following pages, we show how this is done.

How sensitive is the climate?
Evidence from past centuries

Another way to estimate climate sensitivity is to study responses to changes in the natural factors governing climate in previous centuries. Using information from climate proxy data (◀ p.41), such as tree rings and ice cores, scientists estimate how the average temperature of the northern hemisphere varied during the past centuries. They also estimate how the natural factors influencing Earth's climate changed over this time frame, and then compare the two sets of data. Of course, it is important to note that all these estimates come with substantial uncertainties.

Information about how factors that govern climate have changed takes many forms. Sunspot records are available from the early 17th century through modern times. Chemical substances that fall to the surface in snow and become trapped in ice cores can be used to track solar activity even further back in time. Explosive volcanic eruptions can be documented through analyses of the aerosol deposits they left behind in ice cores. The long-term increase in greenhouse gas concentrations since the advent of industrialization is documented in the content of air bubbles trapped within the ice (◀ p.30).

Consistent sensitivity estimates

The equilibrium climate sensitivity estimate of 2–3°C (3.6–5.4°F) derived from proxy data is similar to estimates produced using the modern record (◀ p.84) and geological data (▶ p.88).

Sunspot record
Measurements of sunspots date back to the early 1600s, when they were first recorded with telescopes by European astronomers, such as Galileo. Modern satellite measurements demonstrate that the Sun is slightly brighter in years with high sunspot counts.

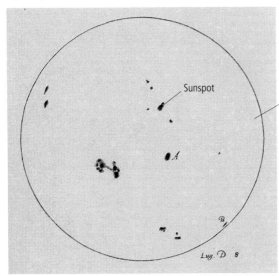

Sunspot

Hand-drawn image of sunspots by Galileo (early 17th century)

Modern image of Sun with sunspots visible

NORTHERN HEMISPHERE TEMPERATURE CHANGES OVER THE PAST SEVEN CENTURIES: SIMULATED VS. ESTIMATES FROM PROXY DATA

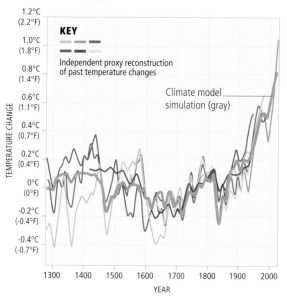

KEY

Independent proxy reconstruction of past temperature changes

Climate model simulation (gray)

Climate scientists compare model predictions with estimated changes in average temperatures in the northern hemisphere derived from proxy data. The proxy temperature estimates match the model simulations well when the assumed equilibrium climate sensitivity is 2–3°C (3.6–5.4°F), meaning that a doubling of atmospheric CO_2 concentrations will lead to a roughly 2–3°C (3.6–5.4°F) warming of the globe.

ESTIMATES OF NATURAL AND HUMAN IMPACTS ON CLIMATE OVER THE PAST MILLENNIUM

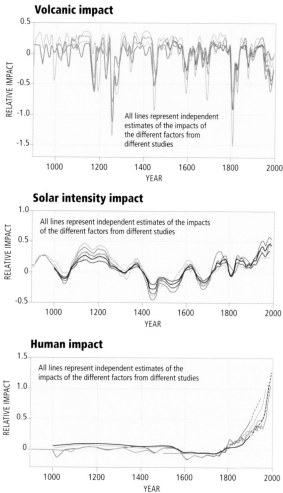

Volcanic impact

All lines represent independent estimates of the impacts of the different factors from different studies

Solar intensity impact

All lines represent independent estimates of the impacts of the different factors from different studies

Human impact

All lines represent independent estimates of the impacts of the different factors from different studies

Scientists drive climate models with the estimated impacts of both natural and human forces. Individual volcanic eruptions have a significant short-term impact, but collectively volcanoes can drive long-term changes when their frequency and magnitude change from one century to the next. Fluctuations in solar output take place on decadal to centennial timescales. Human-caused greenhouse gases have ramped up dramatically over the past two centuries.

Sunspot

How sensitive is the climate?
Evidence from deep time

As we've seen, studies of climate change over the last few centuries can provide us with reliable estimates of climate sensitivity. These sensitivities correspond to changes in atmospheric CO_2, ranging from the pre-industrial level of 280 ppm to the 2014 value of nearly 400 ppm. While significant, this range doesn't include the known glacial–interglacial variations in atmospheric CO_2 over the last 650,000 years: at the peak of the glacial periods, atmospheric CO_2 dipped to 180 ppm. The range also comes well short of possible future increases in atmospheric CO_2, which are predicted to reach nearly 2000 ppm. So how do we determine how climate will respond to the significantly elevated levels of atmospheric carbon dioxide anticipated for the future? We need to look to the ancient past for clues.

Clues from deep time
Geologists estimate that ancient atmospheres contained as much as 1500 ppm of CO_2, and even more (◀ p.40). Therefore, studies of ancient climates can provide important information on climate sensitivity for much larger CO_2 ranges.

A turbulent past
For the last 2 million years, Earth has been swinging in and out of glacial conditions, driven by subtle changes in Earth's orbit around the Sun that are amplified by feedbacks in the carbon cycle and climate system. Data from ice cores demonstrate that fluctuations in CO_2 and temperature have gone hand in hand for at least the last 650,000 years. Feedback loops in the carbon cycle make the question of whether CO_2 is driving climate changes or vice versa virtually impossible to answer. Nevertheless, computer models only simulate the observed cooling when input with low atmospheric CO_2 levels.

Then and now
To learn more about how climate responds to different CO_2 levels, let's step back in time to the height of the last ice age (the "Last Glacial Maximum," or LGM) 21,000 years ago. With much less CO_2 in the atmosphere, the world was then quite

Sea level today

During the last ice age, sea level was 120 m (395 ft) lower

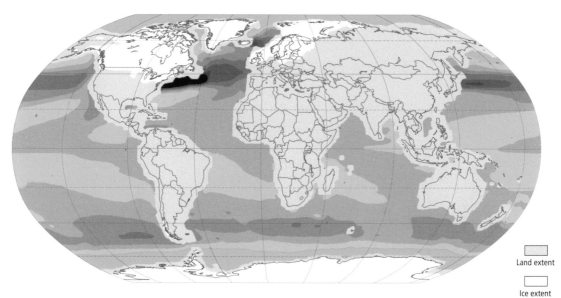

Land extent

Ice extent

HOW MUCH COLDER WAS IT 21,000 YEARS AGO?

Temperature differences between the Last Glacial Maximum, 21,000 years ago, and today, show that the LGM was generally cooler. Mid-to-high latitudes experienced more intense cooling (dark blue), especially near the ice sheets (shown in white) except in the polar seas, where cooling was less intense (green).

°C <	-7.0	-6.0	-5.0	-4.0	-3.0	-2.0	-1.0	0	1.0
°F <	-12.6	-10.8	-9.0	-7.2	-5.4	-3.6	-1.8	0	1.8

SEA SURFACE TEMPERATURE CHANGE (°C/°F) FOR THE LAST GLACIAL MAXIMUM CLIMATE (APPROXIMATELY 21,000 YEARS AGO) RELATIVE TO THE PRE-INDUSTRIAL (1750) CLIMATE.

a different place. The sea level was 120 m (395 ft) lower, because evaporated seawater had fallen as snow and formed the vast ice sheets of the northern hemisphere. A stroll to the beach from Atlantic City, New Jersey, would have taken days, since the shoreline was 80 km (50 miles) east of where it is today. Based on ice-core gas analyses (◀ p.30), we know that the atmosphere's CO_2 content was less than 50% of what it is now. There are other differences between the LGM and today:

- Atmospheric methane was about one-fifth and nitrous oxide was about two-thirds of what they are today.
- Vast ice sheets covered much of Canada, the northernmost U.S., Scandinavia, and northern Europe. These ice sheets were considerably more reflective than the surfaces they replaced. This accounts for half of the cooling, since the ice sheets were reflecting heat rather than absorbing it.

- Earth's orbital configuration was different than it is today (◀ p.66). Because of this, the amount of summer sunshine at high northern latitudes was reduced, so snow from the winter survived the summer and additional ice accumulated.

What can the Last Glacial Maximum teach us about tomorrow's climate?

(Cont.)

CO_2, the LGM, and today

Climate scientists have taken on the challenge of assessing the observed climate of the Last Glacial Maximum. They want to know if the way climate behaved then in response to changes in CO_2 can help us understand how it will behave in the future. The compiled data from the previous page indicate that the global average temperature at the LGM was 3–8°C (5–15°F) cooler than it is today and atmospheric CO_2 levels were much lower as well (180 ppm). The equilibrium climate sensitivity estimate for the LGM is 1–5°C (2–9°F), encompassing the range of our other estimates (◀ p.86). This tells us that data from the LGM confirm our predictions for how the climate will respond to a doubling of atmospheric CO_2.

Ancient data

How will climate respond to even higher levels of CO_2 than those experienced in the glacial–interglacial fluctuations? To answer that question, we must venture much further back into Earth's history. Because ice cores do not go back this far, there is no direct measure of atmospheric composition. So geologists have developed a variety of proxy methods to study atmospheric CO_2 levels. Each tells a somewhat different story, but the overall trends are consistent. Atmospheric CO_2 levels were high 500 million years ago, then fell, reaching a minimum 300 million years ago at the height of the great Permo-Carboniferous glaciations on the supercontinent Gondwana (comprising the modern southern hemisphere continents, as well as the Indian subcontinent and the Arabian Peninsula). After that, levels rose and fell, but reached another maximum about 175 million years ago in the late Triassic. They stayed relatively high through much of the next 100 million years. This was the age of the dinosaurs, when crocodile-like reptiles ventured above the Arctic Circle. Since then, atmospheric CO_2 levels have fallen, reaching another minimum very close to the present day. When combined with the actual paleotemperature estimates, these proxy CO_2 data provide a specific estimate of equilibrium climate sensitivity of 2–5°C (4–9°F) for each doubling of atmospheric CO_2, which is entirely consistent with data from the LGM and with the predictions from state-of-the-art climate models. Most importantly, this analysis precludes a weak equilibrium climate sensitivity (less than 1.5°C, or 3°F, per doubling). This confirms the notion that substantial greenhouse warming is the expected consequence of a buildup of atmospheric CO_2.

Climate sensitivity

As we've seen in the last several pages, scientists have assessed the sensitivity of the climate system to changes in atmospheric CO_2 from a wide variety of sources, from the sediment record of ancient climate shifts to the instrumental record of the last 150 years. Although

A billion years of Earth history
Strata in the Grand Canyon contain evidence of large swings in climate that geologists can relate to corresponding fluctuations in atmospheric CO_2 levels.

uncertainty exists in each of these reconstructions, the amazing result is that a climate sensitivity, expressed as the change in global average surface temperature, of approximately 3°C (5°F) per doubling of atmospheric CO_2 consistently emerges. Somewhat smaller sensitivities are reflected in the historical record and larger sensitivities in the deep time record, suggesting that there may not be one number that characterizes the climate system under all conditions. Nevertheless, there is no indication that the sensitivity is negative, or insignificantly small. In other words, one can state with great confidence that CO_2 buildup has caused, and will continue to cause, significant warming of the planet.

ESTIMATES OF EQUILIBRIUM CLIMATE SENSITIVITY

Estimates of equilibrium climate sensitivity (ECS) and their uncertainties, based on various lines of evidence. Note that the most likely value based on all lines of evidence is very close to 3.0°C (5.4°F).

Likely value   Very likely   Statistical outliers
                                          Historical temperature record   (dashed line)

Most likely   Model simulations of today's average climate

                                          Model simulations of response to CO_2 increase

                                          Paleoclimate data spanning past centuries

                                          Climate response to large volcanic eruptions

                                          Paleoclimate data from last ice age

                                          Model simulations of the last ice age

                                          Paleoclimate data spanning past million years

Overall uncertainty   Paleoclimate data spanning last 400 million years
range based on all
lines of evidence   Assessment of expert viewpoints

                                          Combination of all lines of evidence

| °C | 0 | 1.0 | 2.0 | 3.0 | 4.0 | 5.0 | 6.0 | 7.0 | 8.0 | 9.0 | 10.0 |
| °F | 0 | 1.8 | 3.6 | 5.4 | 7.2 | 9.0 | 10.8 | 12.6 | 14.4 | 16.2 | 18.0 |

EQUILIBRIUM CLIMATE SENSITIVITY

Fossil-fuel emissions scenarios
Predicting the possibilities

Lurking beneath all our predictions about future climate change is a vexing uncertainty: How will human consumption of fossil fuels and land-use practices evolve over the next decades and centuries? The driving forces for consumption are highly complex, involving population growth and per-capita energy demands. These factors are, in turn, closely linked to economic growth and technological advances that can both accelerate consumption and also shift it to other climate-neutral sources.

Possible scenarios

In an attempt to learn more about an uncertain future, climate researchers create and evaluate a range of scenarios for greenhouse gas emissions. This exercise helps them to determine the scope of consequences for a variety of possible future fuel-use scenarios. Initially, these were divided into "business-as-usual" scenarios, which assume ever-increasing rates of fossil fuel use, and "conservation" scenarios, which assume some future reduction of use. For the 4th IPCC assessment, experts from around the world developed four basic "storylines," each representing a group of emissions scenarios for the future. The 5th assessment by the IPCC now defines four scenarios, called Representative Concentration Pathways, or RCPs,

POSSIBLE SCENARIOS FOR THE FUTURE

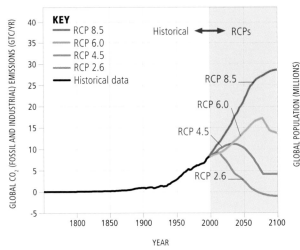

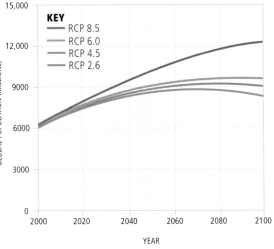

Carbon dioxide emissions
Future emissions differ quite dramatically among the scenarios. The largest growth and cumulative release of CO_2 is associated with the RCP 8.5 fossil-fuel-intensive scenario, while the smallest is associated with the RCP 2.6 scenario.

Global population
In all pathways, global population levels off or starts to decline by 2100; the highest world population (12 billion) is achieved by 2100 in RCP 8.5.

based on their "total radiative forcing" (in watts per square meter) by the year 2100: RCP 2.6, RCP 4.5, RCP 6.0, and RCP 8.5. These four scenarios represent a range of 21st century policies, from the RCP 2.6, which requires strong efforts at mitigation to reduce greenhouse gas concentrations before 2100, to RCP 8.5, which allows the greenhouse gas concentrations to rise beyond 2100. RCP 8.5 is the closest to what might be considered "business-as-usual," the consequences of unabated use of fossil fuels.

Each RCP was created by using "integrated assessment models" that integrate not only climate effects but economic, land use, demographic, and energy considerations as well, yielding CO_2-equivalent greenhouse gas concentration curves for the 21st century that were then converted into an emissions trajectory using carbon cycle models.

Which scenario is most likely?

The IPCC scientists made no attempt to estimate likelihoods for any of these possible scenarios actually occurring; the uncertainties are simply too great. However, these projections do give climate modelers and social scientists a reasonable range of emissions scenario options with which to work.

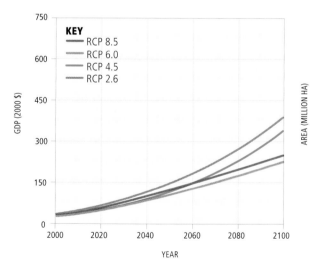

Gross domestic product

Gross domestic product (GDP) increases in all cases, and interestingly, the highest GDP is realized in the RCP 2.6 scenario. Energy consumption increases in all scenarios, with non-fossil-carbon-based energy sources most important in RCP 2.6; RCP 8.5 relies heavily on coal.

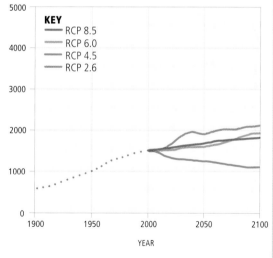

Cropland area

Interestingly, for much of the coming century, cropland area increases most in the two endmember scenarios, RCP 2.6 (because of use of land for bio-energy production) and RCP 8.5 (because of increased demand for crops by a large world population).

The "faux pause"

The fact that temperatures haven't increased as rapidly over the past decade as they did in the prior few decades has led to the false notion that there is a "pause" in global warming.

There *has* been slightly less warming of surface temperatures over this time frame than some climate models predict there should have been, but the discrepancy is smaller in data compilations that better account for data sparsity in the rapidly warming Arctic.

Moreover, the explanations for the lower surface temperatures appear to lie with a number of natural factors that have offset some of the greenhouse warming, including background volcanic activity, a short-term reduction in solar output, and a series of La Niña events, all of which have led to temporary surface cooling not taken into account in most simulations.

By other measures—in particular, the increasing heat content of the oceans, and the accelerating loss of Arctic sea ice—climate change and global warming are unfolding on, or ahead of, schedule.

Less sensitive?

Due in part to the "faux pause" of the past decade, the IPCC dropped the lower end of its estimated range of equilibrium climate

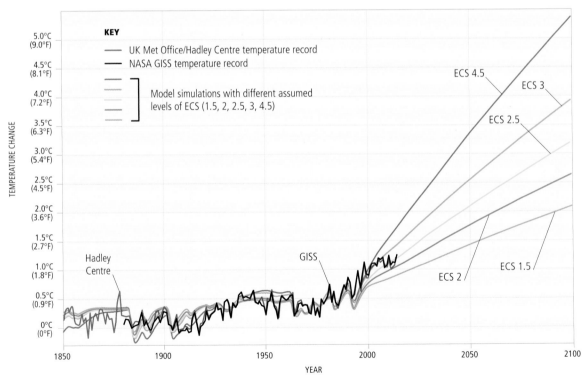

THE FAUX PAUSE
Northern Hemisphere average temperature through 2014 vs. Simulated and Future Projected temperatures based on different assumed levels of Equilibrium Climate Sensitivity or "ECS" (◀pp.86–91).

sensitivity (ECS), by half a degree Celsius in its 5th assessment report, citing a most likely range of 1.5–4.5°C (2.7–8.1°F) eventual warming in response to a doubling of CO_2 concentrations (a level we will cross in a matter of decades given business-as-usual fossil fuel emissions).

Even if ECS were lowered by 0.5°C (0.9°F) from its most likely value, which remains around 3.0°C (5.4°F) based on a variety of lines of evidence (◀ pp.84–91), it would lead to only a modest reduction in warming over the next century.

Perhaps most important, we would still likely cross the threshold of 2.0°C (3.6°F) warming of the globe by mid-century even in this case.

The collective evidence suggests that such an amount of warming would have devastating consequences for society and our environment (▶ pp.120–21).

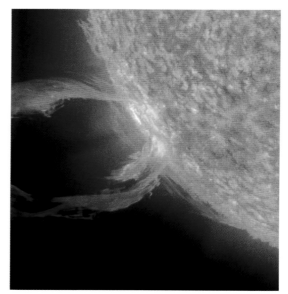

The changing Sun
Changes in solar activity have a modest impact on global temperatures, but potentially larger regional effects. Studies have shown that even small variations in solar radiation can have an impact on climate dynamics.

Deluge or drought
La Niña events can have a significant effect on the North American climate, typically producing above-average rainfall across the northwestern states that leads to flooding. At the same time, states in the southeast and southwest experience below-average rainfall, which encourages the development of much stronger hurricanes out in the Atlantic.

Past IPCC projections
How did they do?

The IPCC has issued future projections for nearly two and a half decades, so we can examine their track record. From the very first assessment report (FAR) in 1990 through the penultimate 4th assessment report (AR4) of 2007, it is reasonable to ask: how well did the future projections compare with what actually happened in subsequent years? The answer is: *quite well!*

The first thing we might examine are the estimated increases in greenhouse gas concentrations, most notably atmospheric CO_2 concentrations. The IPCC made forward projections of CO_2 increases based on a set of scenarios that span the likely range of societal choices

regarding fossil fuel burning and other greenhouse gas-producing activities. The actual CO_2 concentration increases since 1990 have tracked roughly in the center of that range.

Now let's take a look at warming itself. Over the past decade, due to natural factors, the globe has warmed a bit less than the model projected (the temperature changes were projected purely from future increases in greenhouse gases, without taking into account the timing of El Niño events, volcanic eruptions, and other natural factors that influence global temperature). This has led to the false claim by some that "global warming has stopped," but in reality the warming

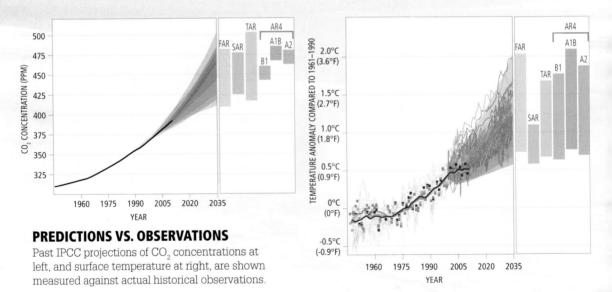

PREDICTIONS VS. OBSERVATIONS
Past IPCC projections of CO_2 concentrations at left, and surface temperature at right, are shown measured against actual historical observations.

KEY

Historical observations

—— AR4 simulation for scenario B1
—— AR4 simulation for scenario A1B
—— AR4 simulation for scenario A2
—— AR4 simulations for 1950–2000

FAR (First assessment report, 1990)
SAR (Second assessment report, 1995)
TAR (Third assessment report, 2001)
AR4 (Fourth assessment report, 2007)

has only been masked temporarily by fleeting natural factors (◀ pp.94–95). Warming is indeed consistent with the history of IPCC-projected warming, if slightly on the low side during the most recent years.

The projections for global sea level rise have fallen firmly within the projected ranges for each of the IPCC assessments. The pace of sea level rise has sped up in recent years as models predicted it would, as greater and greater amounts of land ice begin to melt. As evidence mounts that substantial melting of continental ice sheets has begun earlier than expected, the upper end of the projected range has been revised significantly upward in the most recent assessments (▶ pp.110–111).

Critics sometimes accuse the IPCC of overstating the effects of climate change. If anything, the opposite appears to be true. As noted above, earlier reports appear to have underestimated the rate of likely future sea level rise. But the most striking example comes from the decreasing trend in Arctic Sea Ice. The observed decline in the sea ice left at the end of the summer melting season is greater than the range of AR4 projections. In 2012, the area fell to 3.3 million square kilometers (1.3 million square miles) of ice, a roughly 60% drop relative to mid 20th century levels. IPCC projections indicated that such lows shouldn't be observed until well in the latter half of this century (the model projected rate of decrease is slightly greater in the AR5 model simulations, but the basic discrepancy with observations still exists; ▶ pp.110–111).

GLOBAL MEAN SEA LEVEL RISE

The graph shows past IPCC predictions of global sea level rise measured against actual historical observations.

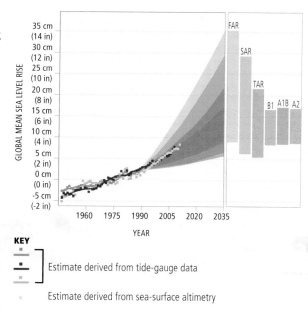

KEY

Estimate derived from tide-gauge data

Estimate derived from sea-surface altimetry

SEA ICE EXTENT

This graph shows past IPCC predictions of Arctic sea ice extent measured against actual historical observations.

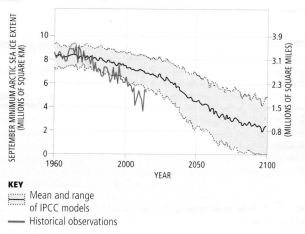

KEY

Mean and range of IPCC models

Historical observations

Melting away

The Arctic has shown a steady decline in the extent of its sea-ice since the late 1970s. Limiting warming to 1–2°C (1.8–3.6°F) by 2100 would help stabilize ice coverage at levels seen today.

The next century
How will the climate change?

Scientists have established a range of possible trajectories for future climate change using climate models (◄ p.68). The spread of these trajectories is due to different possible future greenhouse emissions scenarios, as well as the variances among individual models, which differ in their climate sensitivity (◄ p.92). Typically the results from several different models are averaged to yield a single trajectory for each emissions scenario.

Temperature changes

The predicted increase in global average temperature from 2000 to 2100 is roughly:

- 0.2–1.8°C (0.4–3.2°F) (about 1.2–2.8°C or 2.2–5.0°F relative to pre-industrial time) for the most conservative emissions reduction scenario, involving strong mitigation (RCP 2.6 in figure below).

- 1.1–3.1°C (2.0–5.6°F) (about 2.1–4.1°C or 3.8–7.4°F relative to pre-industrial time) for the mid-range emission scenario (RCP 6.0 in figure below).

- 2.5–4.6°C (4.5–8.3°F) (about 3.8–6.8°C or 6.8–12.2°F relative to pre-industrial time) for the least conservative, i.e., "business-as-usual" scenario (RCP 8.5 in figure below).

These model projections do not take into account some possible positive feedbacks (► p.108) that could further exacerbate global warming. In certain regions, moreover, warming may be considerably greater than the average predicted for the globe as a whole (► p.102).

It is worthwhile noting that even the most conservative scenario yields only even odds of avoiding 2.0°C (3.6°F) warming relative to pre-industrial times. This is the benchmark

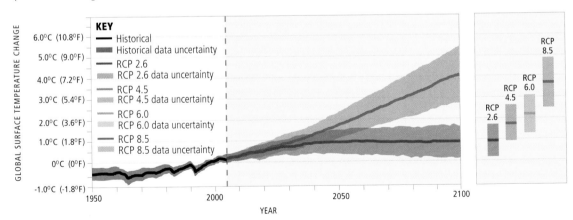

ESTIMATED CO₂ AND TEMPERATURE TRAJECTORIES FOR VARIOUS EMISSIONS SCENARIOS
The graph at left shows the estimated temperature trajectories for various emissions scenarios, as projected by the 44 different state-of-the-art climate model simulations used in the most recent IPCC report. The bold-colored curves show the average predicted warming through the year 2100 temperatures for the two highlighted scenarios (RCP 2.6 and 8.5), with shading to indicate the average spread among models. The bars at right show the range for the projected warming from 2000–2100 for each of the four scenarios shown.

rise that is often cited as constituting dangerous human interference with the climate (◀ pp.94–95).

Precipitation changes

Perhaps of more profound importance than temperature changes are the projected changes in precipitation.

The projected poleward shift in the jet streams of both hemispheres may cause:

- Increased winter precipitation in polar and subpolar regions
- Decreased summer precipitation in many middle latitude regions

Poleward expansion of the tropical Hadley circulation pattern will cause:

- Decreased precipitation in the subtropics

A warmer atmosphere will cause:

- Increased precipitation near the equator.

Hadley circulation pattern

In the Hadley circulation pattern, warm moist air tends to rise, cool, and produce rain near the equator. Depleted of its moisture, it eventually sinks as dry air in the subtropics (◀ p.13 for further details).

Polar cell

Ferrel cell

Air sinks over the subtropical desert zone

Tropical air flows north in this Hadley cell

Dry desert air flows south

Hadley cell

Equator

Warm, moist air rises at the intertropical convergence zone, near the Equator

Hadley cell

Tropical air carries heat south

Air sinks over the subtropical desert zone

Ferrel cell

Polar cell

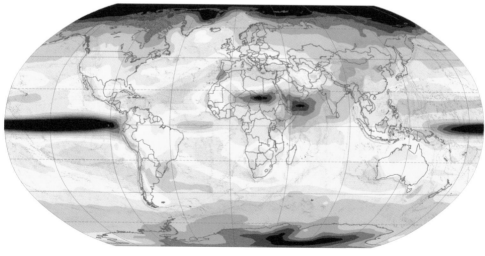

PRECIPITATION PROJECTIONS

Precipitation pattern changes (relative to 1986–2005) projected to occur by 2100 in response to the so-called business-as-usual emissions scenario. Note that there is increased precipitation projected near the equator and in polar and subpolar regions, while subtropical and many mid-latitude regions will likely become drier in summer.

MM < -50 -40 -30 -20 -10 0 10 20 30 40 50 >

IN < -2.0 -1.6 -1.2 -0.8 -0.4 0 0.4 0.8 1.2 1.6 2.0 >

AVERAGE MODEL-PROJECTED CHANGES IN PRECIPITATION (PER DAY) FOR 2081–2100 RELATIVE TO 1986–2005.

(Cont.)

» More drought, more floods

(Cont.)

The combination of *decreased summer precipitation* and *increased evaporation* due to warming surface temperatures is predicted to lead to a greater tendency for drought in many regions. The more vigorous cycling of water through the atmosphere favored by a warming globe will lead to greater rates of both evaporation and precipitation. Consequently, more frequent intense rainfall events and flooding can be expected for many regions as well. Other likely impacts of climate change over the next century include increases in extreme weather phenomena (▶ p.112), and rising sea levels due to melting ice and the warming of the oceans (▶ pp.36–37).

Descending air and high pressure brings warm, dry weather

Southeast trade winds reverse or weaken

Warm water flows eastward, accumulating off South America

Cold upwelling reduced or absent due to weakened trade winds

Low pressure and rising warm, moist air associated with heavy rainfall

El Niño event

During El Niño events, the trade winds in the eastern and central tropical Pacific weaken or even reverse, there is little or no upwelling of cold sub-surface ocean water in the eastern equatorial Pacific, and warm water spreads out over much of the tropical Pacific ocean surface.

Indonesian rainstorm

A man rides his motorcycle with his son through a flooded street in the city of Tangerang, west of Jakarta. On February 13, 2003, heavy rain pelted Jakarta and surrounding cities for five hours, triggering floods in several parts of the area around the capital.

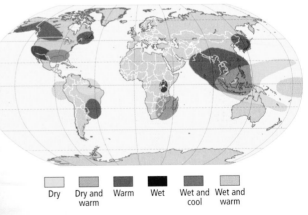

| Dry | Dry and warm | Warm | Wet | Wet and cool | Wet and warm |

LARGE-SCALE IMPACTS OF EL NIÑO (NORTHERN HEMISPHERE WINTER)

El Niño events influence global patterns of temperature and rainfall. The effects of La Niña events are roughly opposite to those shown here for El Niño events.

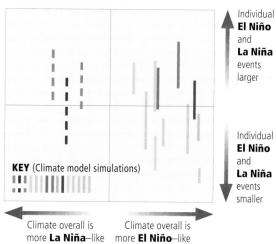

Low-pressure system, positioned farther west than normal

La Niña event
During La Niña events, the trade winds in the eastern and central tropical Pacific are stronger than usual, and there is strong upwelling of cold, deep water in the eastern and central equatorial Pacific.

Pool of warm water positioned farther west than normal

Southeast trade winds stronger than usual

Strong upwelling of cold, deep water

Sea surface cooler than normal in eastern Pacific

El Niño & La Niña

http://goo.gl/aUmngC

El Niño Southern Oscillation (ENSO)

The **El Niño Southern Oscillation (ENSO)** is a natural irregular oscillation of the climate system, involving inter-related changes in ocean surface temperatures, ocean currents, and winds across the equatorial Pacific. This phenomenon alternates every few years between El Niño and La Niña events, which influence weather patterns across the globe.

Uncertain ENSO

Precise regional climate change projections are hampered by uncertainties in how global wind patterns and ocean currents will change. Models don't yet agree on the basic question of whether the climate will become more or less El Niño–like in response to human impacts on climate. Since ENSO is such an important influence on regional patterns of precipitation and temperature, such uncertainties translate to an uncertainty about the patterns of regional climate change themselves. If El Niño events become more frequent, then winter precipitation will increase in regions

such as the desert southwest of the U.S., offsetting any trend toward increased drought in the region (◀ p.52). More El Niño events would also favor worsening drought in regions such as southern Africa.

ENSO VARIABILITY

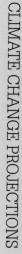

Individual **El Niño** and **La Niña** events larger

Individual **El Niño** and **La Niña** events smaller

KEY (Climate model simulations)

Climate overall is more **La Niña**–like

Climate overall is more **El Niño**–like

Most climate model simulations in the graph above predict a more El Niño–like pattern in response to climate change, but some models predict an opposite, La Niña–like pattern (compare the left and right quadrants). The models are nearly equally split as to whether ENSO variability is likely to increase or decrease in magnitude (compare the upper right and lower right quadrants).

The geographical pattern of future warming

The pattern of projected warming over the planet is far from uniform. The greatest warming will take place over the polar latitudes of the northern hemisphere, due to the positive feedbacks associated with melting sea-ice. Greater warming is projected for land masses than for ocean surfaces, due mostly to the fact that water tends to warm or cool more slowly than land. Accordingly, there is greater warming in the northern hemisphere, which has a higher proportion of land mass, than the ocean-dominated southern hemisphere. Some of the regional variation in warming is due to changes in wind patterns and ocean currents that are also produced by the changing climate. Relatively little warming, for example, is projected to take place over an area of the North Atlantic ocean just south of Greenland, because weakening ocean currents and shifts in the pattern of the northern hemisphere jet stream favor a greater tendency for cold-air outbreaks in this region.

Breaking down the projected pattern of warming at continental scales, it is clear that North America has the potential to see the greatest warming, while South America and Australia are likely to see more modest warming. It should be kept in mind, however, that precise regional temperature projections are limited by uncertainties in how the El Niño Southern Oscillation phenomenon and other regional atmospheric circulation patterns will be affected by climate change (◀ p.100).

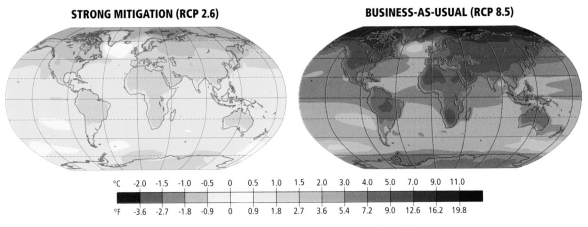

STRONG MITIGATION (RCP 2.6)　　　　　**BUSINESS-AS-USUAL (RCP 8.5)**

°C -2.0 -1.5 -1.0 -0.5 0 0.5 1.0 1.5 2.0 3.0 4.0 5.0 7.0 9.0 11.0

°F -3.6 -2.7 -1.8 -0.9 0 0.9 1.8 2.7 3.6 5.4 7.2 9.0 12.6 16.2 19.8

CHANGE IN AVERAGE SURFACE TEMPERATURE (1986–2005 TO 2081–2100)

MODEL-PROJECTED WARMING BY END OF THIS CENTURY

This graph shows projected surface temperature changes relative to the average temperatures during the late 20th century for scenarios of both strong mitigation and business-as-usual carbon emissions (◀ p.92).

Melting ice field
This image shows summer sea ice breaking up
in the Mackenzie River and the Beaufort Sea.
Warm water from the river greatly accelerates
the melting of the coastal sea ice.

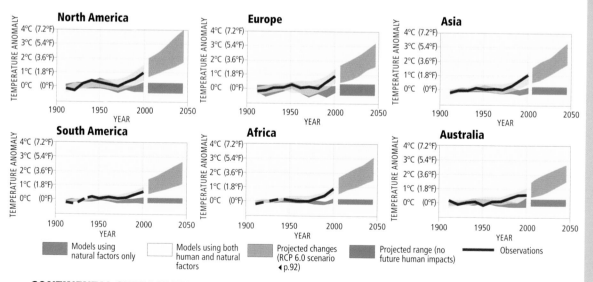

Models using natural factors only	Models using both human and natural factors	Projected changes (RCP 6.0 scenario ◀ p.92)	Projected range (no future human impacts)	Observations	

CONTINENTAL SURFACE TEMPERATURE ANOMALIES: OBSERVATIONS AND PROJECTIONS

The projected future warming given mid-range fossil fuel emissions for each continent is well beyond
the range of temperature changes seen over the past century.

Tipping points, irreversibility, and abrupt climate change

Imagine yourself back in grade school. Your teacher is sitting on one end of a seesaw, and one by one, you and your classmates climb onto the other end. At first nothing happens; your teacher sits comfortably on the ground, the students suspended high in the air. The teacher knows that a tipping point exists, but she would be hard-pressed to predict when it will happen. Of course, when that last student climbs on, the teacher rises into the air and the students sink to the ground.

Now consider the fate of Earth's polar ice sheets in the face of ever increasing atmospheric carbon dioxide levels. While we continue to burn fossil fuels, the ice sheets remain—but we know that at some point the climate will tip into an ice-free state. However, there can be early warning signs of the tipping point, and in fact observations of both the Antarctic and Greenland ice sheets indicate that they are beginning to shrink. Recent studies indicate that the situation in West Antarctica may have already passed the point of no return because the edge of the ice sheet has retreated from a sea floor ridge upon which it had been resting, preventing the flow of glacial ice to the sea. The ice margin is now floating in deep water, and without the friction at its base, ice flow to the ocean melting zone is accelerating.

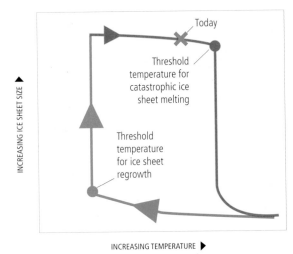

Today

Threshold temperature for catastrophic ice sheet melting

Threshold temperature for ice sheet regrowth

INCREASING ICE SHEET SIZE ▲

INCREASING TEMPERATURE ▶

ICE-SHEET SIZE

HYSTERESIS LOOP
Some environmental systems operate on a hysteresis loop, in which there are two stable states. At low temperatures, the rate of ice sheet growth will be greater than any melting. If the climate warms too much, melting will become unstoppable until temperatures drop substantially.

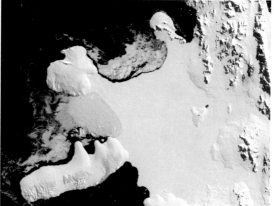

BREAKING AWAY

Parts of the Antarctic ice shelf are showing rapid disintegration as a result of climate change. The image at left shows the Wilkins Ice Shelf on February 28, 2008, just before part of it collapsed. The image at right shows the shelf again on March 17, and the plume of icebergs (pale blue area at the left of the image) that subsequently drifted into the sea.

Ice sheet collapse is not the only potential tipping point in the Earth system. Climate models and observations from the geologic record indicate that the conveyor-belt-like mixing of the Atlantic ocean that brings warm waters to polar latitudes and provides oxygen to the deep ocean may abruptly cease. Of equal concern is the possibility that tropical rainforests may be stressed beyond recovery under warmer global climate conditions

One alarming aspect of tipping point behavior is that it can exhibit a phenomenon known as hysteresis: the path back to the original condition can be much different than the path that led the system to the altered state. For example, the collapse of the ice sheets could be catastrophic once a threshold temperature is reached. The regrowth of those same ice sheets, however, will only occur once Earth's temperature has fallen well below the present temperature (see graph)—and then the process could take millennia.

On societal timescales, and unlike the teacher suspended in the air, the ice sheet collapse (and attendant massive sea-level rise) could be very long-lasting.

Critical mass

Predicting when an environmental system will reach its tipping point is more difficult than the simple physics of a seesaw. Although there may be signs that something is happening, it may be hard to determine which factors are responsible until it is too late to initiate a remedy.

Carbon-cycle feedbacks
Nature's response to CO$_2$

You might think that all of the carbon dioxide (CO$_2$) released through fossil-fuel burning and deforestation has simply accumulated in the atmosphere. Yet detailed analysis shows that 55% of the CO$_2$ we've pumped into the atmosphere since 1750 has "disappeared." Actually, scientists know where it went. Much of the "missing" CO$_2$ has dissolved into the ocean, and the rest has been stripped from the atmosphere via photosynthesis and incorporated into living biomass. (Photosynthesis is the process by which plants, and a few other organisms, use energy from sunlight to convert CO$_2$ into sugar—the "fuel" used by all living things on Earth.) The 45% of the CO$_2$ that has not "disappeared" but accumulated in the atmosphere is termed the "airborne fraction." In effect, nature has already responded to fossil-fuel burning to a certain degree, and somewhat reduced the human impact on atmospheric composition and climate. But nature has its limits and humans are beginning to push up against them.

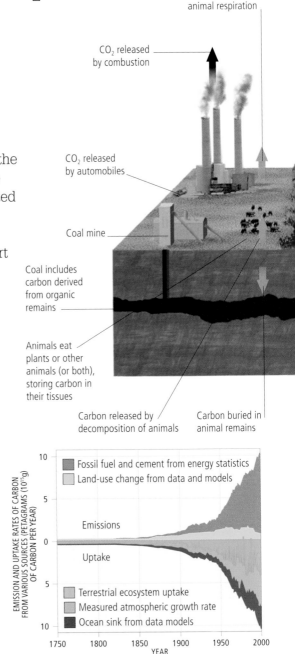

CO$_2$ released by animal respiration

CO$_2$ released by combustion

CO$_2$ released by automobiles

Coal mine

Coal includes carbon derived from organic remains

Animals eat plants or other animals (or both), storing carbon in their tissues

Carbon released by decomposition of animals

Carbon buried in animal remains

WHERE DID ALL THE CO$_2$ GO?

EMISSION RATE OF CARBON FROM FOSSIL FUEL COMBUSTION AND CEMENT MANUFACTURING (PETAGRAMS (10^{15}g) OF CARBON PER YEAR)

Cement
Gas
Oil
Coal

YEAR

EMISSION AND UPTAKE RATES OF CARBON FROM VARIOUS SOURCES (PETAGRAMS (10^{15}g) OF CARBON PER YEAR)

Fossil fuel and cement from energy statistics
Land-use change from data and models

Emissions

Uptake

Terrestrial ecosystem uptake
Measured atmospheric growth rate
Ocean sink from data models

YEAR

Emissions and uptake rates in petagrams (10^{15}g) of carbon (C) per year. The graph at left shows C emissions from fossil-fuel combustion and cement manufacturing. The graph at right shows the sum of these along with emissions from deforestation and other land-use changes and uptake in terrestrial ecosystems, the atmosphere, and the oceans.

THE CARBON CYCLE

The main reservoirs of carbon are the atmosphere, the ocean, and vegetation, soils, and detritus on land. Marine life represents a very small carbon reservoir. On multi-millennial time scales, geologic reservoirs also become important. Various processes transfer carbon between these reservoirs, including photosynthesis and respiration, ocean–atmosphere gas exchange, and ocean mixing.

KEY

→ Carbon movement
→ Weathering and erosion
→ Human carbon transformation

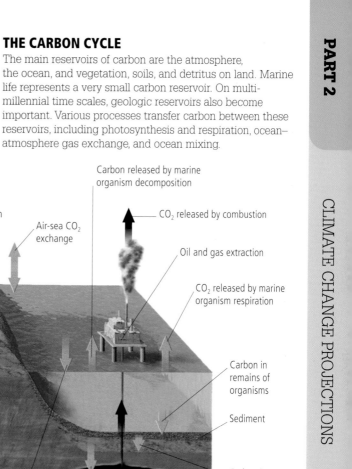

CO_2 released by plant respiration

CO_2 absorbed by photosynthesis

Carbon stored in plant tissues

CO_2 released by volcanic eruption

CO_2 dissolved in water

Rivers carry eroded carbon to the ocean

Air-sea CO_2 exchange

Carbon released by marine organism decomposition

CO_2 released by combustion

Oil and gas extraction

CO_2 released by marine organism respiration

Carbon in remains of organisms

Sediment

Carbon in sediment turns into limestone and organic-rich shale

Carbon released by decomposition of plants

CO_2 in rain weathers rocks

Carbon moves from sediment to oil and gas

Oil and gas

Carbon buried in plant remains

CO_2 absorbed by photosynthesis

The Ocean's Green Machines

http://goo.gl/Xqfjq1

Phytoplankton

Marine phytoplankton like these diatoms play an important role in transferring fossil-fuel carbon dioxide from the atmosphere into the ocean.

(Cont.)

Positive feedbacks prevail

Despite nature's best efforts to counter our impact on the planet, atmospheric CO_2 levels have nonetheless continued to rise, and the planet is warming. Unfortunately, warming reduces nature's ability to absorb CO_2. A number of feedback loops (◄ p.24) are involved, both positive (enhancing warming) and negative (reducing warming), but positive feedbacks prevail on all but multi-millennial timescales.

Warmer land
Positive feedback:
Soil microorganisms increase their growth and respiration rates as their environment warms. One of the waste products of their metabolism is CO_2. As a result, carbon in soils is now being converted to CO_2 at increasing rates.

Negative feedback:
This release of CO_2 to the atmosphere by soil microorganisms offsets gains made by plants responding favorably in their growth to elevated CO_2 levels (so-called CO_2 fertilization, a negative feedback; ► p.117).

Warmer ocean
Positive feedback:
A warmer ocean has less ability to absorb carbon dioxide, just as an opened can of warm soft drink loses its carbonation and goes flat.

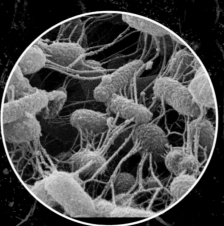

Soil microorganisms
Bacteria, such as these shown here, are microorganisms that decompose organic matter in soils.

Ocean acidification
Negative feedback:
Acidification of the surface ocean (► p.126) reduces the production of calcium carbonate (limestone/$CaCO_3$) by organisms such as corals and tiny plankton. When these organisms grow their $CaCO_3$ skeletons (i.e., to calcify), CO_2 is released to the water. So calcification reduces the ocean's ability to take up fossil-fuel CO_2. However, it is predicted that some calcifying organisms will become extinct this century. If calcifying plankton and corals become less abundant, then the resulting reduction in the rate of calcification will slightly increase the ocean's ability to take up carbon dioxide.

Goings-on down under
Earthworms, bacteria, and fungi consume plant matter buried in soil, releasing CO_2 that diffuses into the atmosphere above.

Pump problems
Positive feedback:
Calcium carbonate is a relatively dense mineral that acts as ballast once an organism dies, carrying its decaying tissue to great depths in the ocean. This "pump" of carbon removes CO_2 from surface waters, allowing more fossil fuel CO_2 to be absorbed. Loss of the ballast via ocean acidification reduces the ocean's ability to take up atmospheric CO_2.

A sluggish ocean
Positive feedback:
A slowing of ocean circulation in response to global warming reduces the mixing up of nutrients at the ocean's surface, which slows biological productivity. This weakens the action of the biological pump, further reducing the ocean's ability to absorb CO_2.

Rock weathering
Negative feedback:
Increased temperatures and rainfall stimulate the chemical weathering of rocks on land (the process that turns rock into soil and into dissolved salts in rivers). Atmospheric carbon dioxide dissolved in rain forms carbonic acid, which aids the rock-weathering process during which CO_2 is converted to other forms of dissolved carbon. Increased weathering, therefore, removes CO_2 from the atmosphere.

Overall effect
Models that simulate both the carbon cycle and climate have been run with some, but not all, of these feedbacks taken into consideration. The overall effect of these feedbacks is a more rapid buildup of atmospheric CO_2, and a warmer climate. This additional warming is reflective of a carbon cycle that is approaching the limit of its

Melting ice and rising sea level

Sea level is predicted to rise with global warming for two reasons. First, water, like most liquids, becomes less dense (i.e., it expands) as it warms. A small rise of 0.1–0.4 m (0.3–1.3 ft) is predicted by 2100, depending on the emissions scenario, due to this effect alone. Second, melting ice is likely to have a major impact on the sea level. It is important to note, however, that not all ice plays an equal role here. The disappearance of high-latitude sea-ice (▶ p.148), while significant in its own right, will not be a contributor. Much as melting ice cubes in a glass of water do not cause the level of water to rise, melting sea-ice will not cause the sea level to rise. On the other hand, melting continental ice will definitely contribute to a sea level rise.

Significant rise

The continental ice resides in two basic forms. First, there are the permanent ice caps and glaciers in mountain ranges at high latitudes, and even at equatorial latitudes (◀ p.64). Melting all of this ice, however, would only add at most a sea level rise of about 0.5 m (1.6 ft).

More significant are the Greenland and Antarctic continental ice sheets. There is evidence that significant melting of the Greenland ice sheet is already underway, but the rate of future melting is difficult to estimate. Model simulations indicate that local warming over Greenland may exceed 4.0°C (7.2°F) by 2100 given business-as-usual greenhouse gas emissions (◀ p.102). Current ice-sheet models indicate that such a warming could lead to the eventual irreversible melting of the Greenland ice sheet, resulting in roughly 5–7 m (16–23 ft) of global sea level rise. Melting of the most unstable part of the Antarctic ice sheet (the West Antarctic ice sheet) could add an additional 5 m (16 ft). Some recent studies indicate that melting of ice from below by the warming ocean may already have committed us to much of that already, though just how long it would take is a matter of uncertainty (likely at least two centuries, but possibly sooner).

Even faster than we thought

But even state-of-the-art models do not account for some newly observed effects that scientists now believe could significantly accelerate the rate of melting. For example, recently it has been discovered that crevices (called "moulins") are forming in melting continental ice. These moulins allow surface meltwater to penetrate deep into the ice sheet and lubricate the base, allowing large pieces of ice to slide quickly into the ocean. If this phenomenon becomes increasingly widespread, it could lead to a far more rapid disintegration of the ice sheets than predicted by any current models. Scientists were particularly surprised by unprecedented, widespread surface melt that was observed over the Greenland ice sheet in summer 2012.

Moulin in Greenland
Crevices (moulins), such as this one in Greenland, allow surface meltwater to penetrate deep into ice sheets.

Sea level is currently projected to rise between 0.5 and 1.2 m (1.6 and 3.9 ft) by 2100.

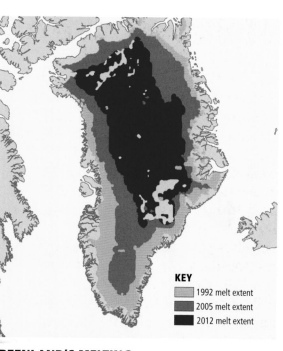

KEY
- 1992 melt extent
- 2005 melt extent
- 2012 melt extent

GREENLAND'S MELTING CONTINENTAL ICE SHEET

This map shows changes in the extent of the region of summer melting in Greenland.

PROJECTED SEA LEVEL RISE

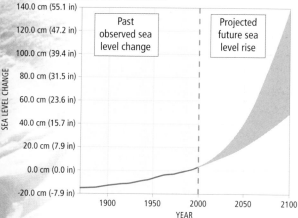

By relying on observations of the past relationships between global temperature and rates of sea level rise, and global temperature projections for the next 100 years (◀ p.98), scientists can make projections about future sea level rise. The projected rise by 2100 is between 0.5 and 1.2 m (1.6 and 3.9 ft), depending on the prevailing emissions scenario (◀ p.92).

Future changes in extreme weather

As climate changes, it is likely that the frequency and intensity of extreme weather events will change. For certain extreme weather events, such as severe frosts and extended heat waves, the science is fairly definitive and the predicted changes are intuitive. The greater the amount of warming, the more pronounced these trends will be. Changes are predicted to vary regionally.

New trends are expected to emerge for various types of extreme weather, including heat waves, heavy downpours, and frosts. In the maps above and on the following pages, the colors represent a relative scale. Variations within color fields indicate regions where climate models predict a significant increase or decrease in the quantity in question.

Fewer frosty days

- As temperatures warm, the probability of frosts (nights when temperatures dip below freezing) will decrease markedly.

- The greatest decrease in frost days is likely to occur in regions such as interior North America and Asia, where winter temperatures have been traditionally the coldest.

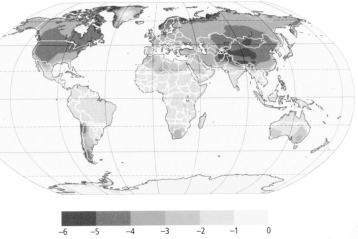

−6   −5   −4   −3   −2   −1   0

CHANGES IN FROST DAY FREQUENCY

This map shows changes in the occurrence of frost days projected by the late 21st century (2080–2099) relative to the observed frequency of occurrence in the late 20th century (1980–1999). A relative scale is used, where a single unit ("1") represents a shift as large as the typical range of year-to-year variations. The term "frost days" refers to the number of days in the year when the minimum nightly temperature drops below freezing.

Hard frost
Familiar cold-weather scenes, like this one in Colorado, will become less common as our climate warms.

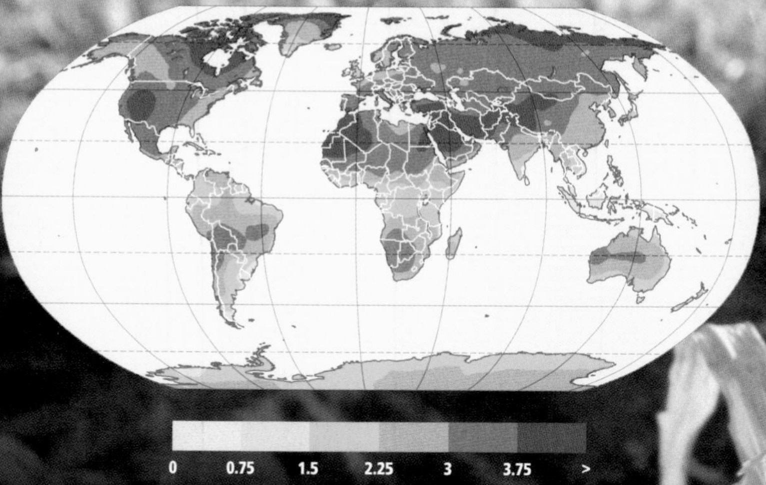

| 0 | 0.75 | 1.5 | 2.25 | 3 | 3.75 | > |

CHANGES IN HEAT WAVE FREQUENCY

This map shows changes in the occurrence of heat waves projected by the late 21st century (2080–2099) relative to the observed frequency of occurrence in the late 20th century (1980–1999). A relative scale is used, where a single unit ("1") represents a shift as large as the typical range of year-to-year variations. A "heat wave" is a minimum of five consecutive days when the high temperature is at least 5˚C (9˚F) above the average.

Dry corn
An increase in the frequency of blistering heat waves associated with climate change may destroy crops like the corn in this parched field.

More heat waves

▪ Heat waves (very high temperatures sustained over a number of days) are likely to become more intense, more frequent, and longer lasting.

▪ The greatest increase in heat waves is predicted to occur in areas such as the western U.S., North Africa, and the Middle East, where feedback loops associated with decreased soil moisture may intensify summer warmth (◀ p.100).

(Cont.)

113

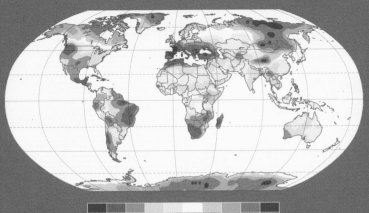

Wet days and dry days

Most model simulations also indicate that increases are to be expected in the frequency of very intense precipitation events and corresponding flooding. These changes are due to the more vigorous water cycle that will accompany a warmer climate, with greater amounts of moisture in a warmer atmosphere (◀ p.24) leading to more intense precipitation events. Seemingly paradoxical, while many regions are likely to become drier, scientists predict that even in those regions individual rainfall or snowfall events will become more intense, although longer dry spells will separate them. Also counter-intuitively, individual

Roadway under water

Here, a vehicle has been partially submerged due to flooding of the Schuylkill River in Philadelphia, Pennsylvania, in May 2014. During a nine-hour period, Philadelphia received about 125 mm (almost 5 in) of rain and nearby counties received even more, causing widespread river flooding. As the atmosphere warms and can hold more water, heavy rainfall events like this are predicted to become more common.

| −1.25 | −1 | −0.75 | −0.5 | −0.25 | 0 | 0.25 | 0.5 | 0.75 | 1 | 1.25 |

CHANGES IN PRECIPITATION INTENSITY

Projected changes in the pattern of precipitation intensity by the late 21st century (2080–2099) relative to the observed frequency of occurrence in the late 20th century (1980–1999). A relative scale is used, where a single unit ("1") represents a shift as large as the typical range of year-to-year variations.

| −1.25 | −1 | −0.75 | −0.5 | −0.25 | 0 | 0.25 | 0.5 | 0.75 | 1 | 1.25 |

CHANGES IN MAXIMUM DRY SPELL LENGTH

Projected changes in the occurrence of changes in maximum dry spell length by the late 21st century (2080–2099) relative to the observed frequency of occurrence in the late 20th century (1980–1999). A relative scale is used, where a single unit ("1") represents a shift as large as the typical range of year-to-year variations.

winter snowfall events may become heavier in regions that remain cold enough for winter snow but are nonetheless warmer (and therefore more moisture-laden) as a result of climate change.

Severe storms

It is more difficult to determine how extreme weather events such as tornados, severe thunderstorms, and hailstorms will change. This is because those events involve processes that occur at too small a scale to be reproduced in most model simulations. However, it is likely that even if such events do not become more severe or more common in general, individual storms will likely be associated with more severe downpours and more frequent flooding due to the greater amount of water vapor that a warmer atmosphere can hold.

Hurricanes and cyclones

But what about hurricanes (and their weaker cousins, tropical cyclones)? We know that there has been a recent trend toward more intense hurricanes in certain basins, such as the tropical Atlantic basin, and that these trends closely mirror warming ocean surface temperatures (◀ p.62). A warmer ocean surface, all other things being equal, is likely to fuel more intense tropical cyclones, with stronger sustained winds. The combination of sea level rise and stronger tropical storms could pose a "double whammy" when it comes to damaging storm surges. Model simulations indicate a likely shift toward the strongest (Category 4 and 5) tropical cyclones over the next century, given projected climate changes. There are some important unanswered questions, however. For example, we know that El Niño events change wind patterns over the tropical Atlantic region in such a way as to create unfavorable conditions for tropical cyclone formation. And there is still uncertainty about how El Niño will change in response to climate change (◀ p.100). Climate scientists are currently working to resolve such open questions.

Storm surge
Hurricane Sandy hit the coast of New Jersey on October 29, 2012. Sea levels in some areas along the east coast of the U.S. are thought to have risen by up to 30 cm (12 in) over the last 100 years as a result of global warming.

Stabilizing atmospheric CO_2
Is a greenhouse world a better world?

The rapid rise of atmospheric carbon dioxide (CO_2) levels over the last two centuries is a clear outcome of society's hunger for cheap energy—a hunger that is insatiable and is being fed an unhealthy diet of fossil fuels. Recognizing this, scientists have been studying future fossil-fuel-use scenarios with substantially elevated atmospheric CO_2 levels. The IPCC Fifth Assessment Report paid specific attention to emission scenarios that stabilized atmospheric CO_2 levels at 550 ppm (RCP 4.5), 750 ppm (RCP 6.0), and 2000 ppm (RCP 8.5) (◀ p.92 for a description of the RCP scenarios) sometime after 2100: these values are considerably higher than the 2014 value of about 400 ppm. The more optimistic scenario (RCP 2.6) had CO_2 levels peaking in the mid 21st century at about 450 ppm and then declining slowly thereafter.

To prevent atmospheric CO_2 levels from exceeding 450 ppm, fossil fuel use must peak by 2020

Even with atmospheric CO_2 levels peaking at 450 ppm, global temperature increases an additional 1°C (1.8°F) during the 21st century, or 2°C (3.6°F) since pre-industrial times, and sea level rises by 0.5 m (20 in) or more by 2100

GREENHOUSE GAS LEVELS RESULTING FROM VARIOUS EMISSIONS SCENARIOS

The emission scenarios on these graphs show different possible stabilization levels of atmospheric CO_2 and other greenhouse gases over the next three centuries. The colors of the lines in these graphs correspond to the colors that represent the emissions scenarios described on p.92. ECPs refer to the time period after 2100, the "extended" concentration pathways in the graphs below.

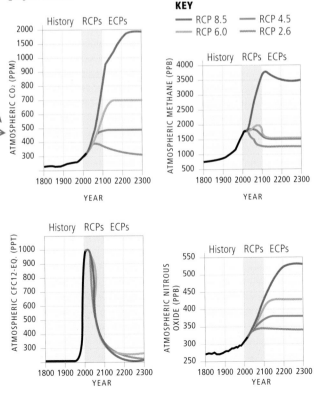

KEY

— RCP 8.5   — RCP 4.5
— RCP 6.0   — RCP 2.6

The need to act now

A broad range of CO_2 stabilization targets has been studied, ranging from 550 ppm to 2000 ppm. The top left graph on the facing page shows the gradual climb to these levels over the next three centuries.Some interesting characteristics of these curves emerge:

■ The lower the stabilization target, the sooner peak emissions of fossil fuel carbon dioxide must occur. In other words, the lower the level at which we want to stabilize the CO_2 levels in the air, the sooner we have to cut back on fossil fuel use. To stabilize atmospheric CO_2 levels at 550 ppm, we would need to reach peak usage before 2040. To stabilize at 750 ppm, we need to reach peak usage by 2080.

■ Lower stabilization levels can be achieved only with lower peak emissions; while the 750 ppm target allows CO_2 emission rates to double, the 550 ppm target allows them to increase by only 50% or so.

■ All stabilization targets require sharp reductions in CO_2 emissions following the peak. Low stabilization targets require that emission rates fall below the current rate within a few decades.

It is also interesting to note that the projected climate changes associated with even the most conservation-minded emission targets are substantial. The long-term warming projected for RCP 2.6 is about 1°C (1.8°F), during the 21st century, or 2°C (3.6°F) since pre-industrial times (◄ p.98), an increase that could lead to a sea level rise of half a meter (20 in) or more (◄ p.110) by 2100. In comparison, the RCP 4.5 scenario stabilizes at an even higher CO_2 level (about 550 ppm) by the year 2100. Moreover, even with the CO_2 level stabilized, the temperature and sea level will continue to rise as the sluggish climate system adjusts to the new atmospheric composition, committing us to continued warming and coastal inundation for decades to come.

Will more CO_2 benefit plants?

Given the serious reductions that would be required to achieve low stabilization targets, some people argue that the higher CO_2 stabilization targets shouldn't be considered failures, but rather desirable objectives for beneficial climate modification. This line of reasoning argues that plants require CO_2 for photosynthesis and that more CO_2 should benefit plants (CO_2 fertilization), including the crops that feed the people of the world. Crops grown under ideal conditions in greenhouses with elevated carbon dioxide levels do outperform those grown under ambient atmospheric conditions. In the presence of elevated CO_2, plants do not have to open their pores as wide; this reduces water loss and infection by pathogens, and encourages growth. However, these benefits cannot be fully realized in nature if other factors, such as a lack of nutrients or inadequate soil, are limiting growth.

The existence of growth-limiting factors, combined with other negative impacts from elevated CO_2 levels—such as ocean acidification and loss of coral reefs (▶ p.126)—suggests that the "greening of planet Earth" may not be an achievable or desirable outcome of fossil-fuel burning.

Part 3
The Impacts of Climate Change

Recent studies indicate that the future impacts of climate change are likely to be far more significant than those observed to date. Human societies, natural habitats, and a myriad of animal and plant species will all be affected by changes in temperature and precipitation patterns in the decades ahead. The precise impacts will depend on the rate and amount of warming, and on the adaptive measures taken by society.

Key ideas

Climate change & society

- Climate change puts stress on human societies. If climate change continues, it will likely lead to greater competition for natural resources, threats to food supplies, increased risks to human health, and international conflict.

- The populations of the world's coastal and low-lying regions are at particular risk because they are exposed to temporary flooding and permanent inundation caused by rising sea levels.

- Warming could lead to more atmosphereic pollution by accelerating ozone production and promoting air stagnation.

- In North America, sea-level rise, extreme weather events, and wildfires pose a serious threat to people, infrastructure, and the economy.

Climate change & the environment

- Climate change tests the resilience of natural ecosystems. Some systems might adapt, but for others the amount and rate of change will be too great, leading to ecosystem destruction and a loss of biodiversity.

- Coral reefs are among the most vulnerable ecosystems because they have little scope for adaptation.

- Human intervention in natural systems could be a powerful enough force to cause a mass extinction of animals and other living organisms.

The rising impact of global warming

A world under stress

A list of the potential impacts of global warming on humanity and planet Earth makes for sobering reading. These impacts include a greater tendency toward drought in some regions, the widespread extinction of plant and animal species, decreases in global food production, the loss of coastlines and coastal wetlands, increased storm damage and flooding in

Dry lake bed
During the 2011 Texas drought, Lake Fisher (also known as the O.C. Fisher Reservoir) in San Angelo disappeared, and it has not come back.

many areas, and a wider spread of infectious disease. Stresses such as these could, in turn, lead to increased competition for natural resources, over-taxed social services and infrastructure, and conflict between regions and nations. Sustainable approaches to development will be necessary to decrease the vulnerability of society, ecosystems, and the environment to future changes.

EFFECT OF FURTHER TEMPERATURE CHANGES

GLOBAL WARMING IMPACT SCALE

+4.6°C (8.3°F)

Global economic losses of up to 5% of GDP

At least partial melting of Greenland and West Antarctic ice sheets, resulting in eventual sea-level rises of 5–11 m (16.4–36.1 ft)

+3.6°C (6.5°F)

2100

Substantial burden on health services

Decreases in global food production

About 30% of global coastal wetlands lost

2090

2080

+2.6°C (4.7°F)

2070

Changes in natural systems cause predominantly negative consequences for biodiversity, water, and food supplies

Millions more flood victims every year

2060

Widespread coral mortality

2050

+1.6°C (2.9°F)

2040

Human mortality increases as a result of heat waves, floods, and droughts

9%–31% species extinction

2030

2020

+0.6°C (1.1°F)

AMOUNT OF GLOBAL WARMING

2010

(0.6°C/1.1°F increase over pre-industrial—before 1750s)

INCREASE OVER PRE-INDUSTRIAL LEVELS

• Decreases in water availability; more frequent droughts in many regions

• Wildfire risk increases, as do flood and storm damage

• The burden from increased incidence of malnutrition and diarrheal, cardio-respiratory, and infectious diseases escalates

121

Is it time to sell that beach house?

Nearly 10% of the world's population lives in coastal and low-lying regions, where the elevation is within 10 m (32.8 ft) of sea level. In some places, such as Bangladesh, this population figure is nearer to 50%. Rising sea level (◀p.110), increasing destruction associated with tropical cyclones (◀p.62), increasing coastal erosion, and larger wave heights all pose serious threats to coastal and low-lying regions.

Soggy cities

The most obvious threat associated with global sea level rise is coastal inundation. In North America, for example, significant loss of land on the mid-Atlantic and northeast coastlines could occur with just 6 m (19.6 ft) of sea level rise—an amount that would be all but guaranteed by the melting of the Greenland ice sheet alone (◀p.110). The flooding of the New York metro area due to Hurricane Sandy in October 2012 is estimated to have been increased by roughly 65 square km (25 square miles) due simply to the

SOUTHERN FLORIDA

Southern Florida would be submerged if sea levels were to rise between 4 m (13.1 ft) and 8 m (26.2 ft).

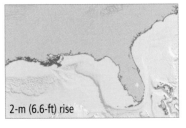

1-m (3.3-ft) rise

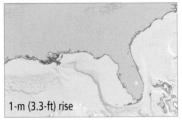

2-m (6.6-ft) rise

4-m (13.1-ft) rise

Sea Level Rise

http://goo.gl/hmv2NP

NORTHEAST COASTLINE

Most of New York City and Boston would be submerged if sea level were to rise by 6 m (19.6 ft).

6-m (19.6-ft) rise

KEY

Land submerged if sea level rises

8-m (26.2-ft) rise

modest 0.3 m (1 foot) of sea level rise that has already taken place. Substantial portions of Europe's "low countries" including Belgium and the Netherlands would also be submerged with a sea level rise of 4–8 m (13.1–26.2 ft).

Human loss

Even those coastal regions not inundated by higher sea levels will be subject to increased exposure to flood and storm damage, more intense coastal surges, and altered patterns of coastal erosion. Associated impacts are likely to include loss of human life, damage to human infrastructure and real estate, degraded water quality, and decreased availability of fresh water due to saltwater intrusion. Coastal habitats will be lost if water levels and wave heights substantially increase. Significant population displacement will also be a factor. Communities, habitats, and economies on all of the major continents will be affected by even just 1 m (3.3 ft) of sea level rise. The cost rises dramatically at 5 m (16.4 ft) and 10 m (32.8 ft). The melting of the Greenland ice sheet would ensure the former, while the additional melting of a substantial piece of the West Antarctic ice sheet would ensure the latter.

GLOBAL LOSSES

These bar graphs show the impacts of sea level rises of 1 m (3.3 ft), 5 m (16.4 ft), and 10 m (32.8 ft). Total global and North American losses are shown. Note that losses are sizeable even in the event of 1 m (3.3 ft) of sea level rise, which could plausibly occur by the end of this century.

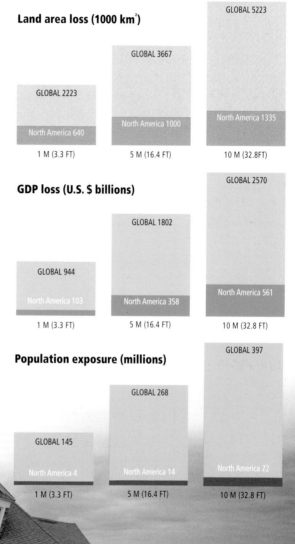

Land area loss (1000 km²)

GLOBAL 5223

GLOBAL 3667

GLOBAL 2223

North America 640

North America 1000

North America 1335

1 M (3.3 FT)   5 M (16.4 FT)   10 M (32.8FT)

GDP loss (U.S. $ billions)

GLOBAL 2570

GLOBAL 1802

GLOBAL 944

North America 103

North America 358

North America 561

1 M (3.3 FT)   5 M (16.4 FT)   10 M (32.8 FT)

Population exposure (millions)

GLOBAL 397

GLOBAL 268

GLOBAL 145

North America 4

North America 14

North America 22

1 M (3.3 FT)   5 M (16.4 FT)   10 M (32.8 FT)

Rough tides
The waves generated by Hurricane Sandy wreaked havoc along the New Jersey coast in October 2012. Homeowners face the dual threat of stronger storms and rising sea level as a result of climate change.

Ecosystems
Worth saving?

Perhaps your drive to work takes you through a reeking, swampy area that's often subject to flooding. Imagine a future in which this lowland quagmire is filled and a new interstate is built that takes you rapidly through the area without the smell and inconvenience. An improvement? In some ways, perhaps. But was anything of value lost when the ecosystem was destroyed? Does the attractive pond the transportation department built to replace the swamp serve the same function as the wetland that was destroyed? This begs a further question: are wetlands and other natural ecosystems of any real use to us? The answer is an overwhelming "Yes"!

What is an ecosystem?

An ecosystem is an interdependent community of plants, animals, and microscopic organisms, and their physical environment. All these different elements interact and form a complex whole, with properties unique to that particular combination of living and non-living elements. Ecosystem boundaries are generally delineated by climate: desert ecosystems in the subtropics, tropical rain forest ecosystems near the equator, and tundra ecosystems near the poles. As climates have changed in the geologic past, ecosystems have shifted in response. But past climate changes were slower than the projected future changes. Will the ecosystems of today be able to adjust their boundaries as the climate changes, or will they be stranded with incompatible climates? And why, for that matter, should it concern us?

Wetland predator
American alligators are vital to the balance of wetland ecosystems—as predators, they maintain prey levels, and their holes remain filled with water during dry seasons, providing refuge for other animals.

Why we need wetlands

Wetlands provide an important service to their surroundings. Storm waters that flow into a wetland lose energy and spread out across a broad area, thus reducing flooding downstream. Furthermore, sediment and contaminants—such as iron and acidity from mine runoff and nitrogen from farm fertilizers—are removed as water percolates through wetlands before entering our drinking and irrigation water. Wetlands are also biologically diverse ecosystems, and provide homes for endangered species and a refuge for migrating birds. This makes them ideal places for hiking, bird watching, canoeing, and fishing.

Worth saving

Ecosystems are valuable to humanity because they assist us with:

- **Provisions:** food (seeds, fruits, game, spices); fiber (wood, textiles); medicinal and cosmetic products (dyes, scents)

- **Environmental regulation:** climate and water regulation; water and air purification; carbon sequestration; protection from natural disasters, disease, and pests

- **Cultural benefits:** appreciation of, and interaction with, the natural world; recreational activities

Ecosystems are reasonably resilient to change, including a modest amount of human disturbance. But there are limits to this resilience, and human activity is already pushing those limits in some ecosystems (▶ p.126). Climate change not only challenges the persistence of ecosystems—it may also lead to the extinction of species that cannot adapt or migrate rapidly enough (▶ p.130).

Unlike local highway construction, the stress placed on ecosystems by climate change is an insidious one—less obvious, but perhaps more permanent. As climate zones (◀ p.76) shift poleward, ecosystems will likely follow. However, they may be stopped from doing so by natural factors (such as incompatible soils) and human development (roads, cities, agriculture). The result could be widespread destruction of ecosystems, with loss of benefits to society and a significant reduction in global biodiversity.

Humans meet nature
"Alligator Alley", the highway through the Florida Everglades wetlands ecosystem, was designed with numerous overpasses to minimize environmental impact and threats to the surrounding ecosystems.

Coral reefs Will ocean acidification be their demise?

Coral reefs are among the world's most diverse ecosystems. On a single snorkeling adventure in a healthy reef you can see more species of animals than you can during a lifetime of hiking in mid-latitude forests. Reefs also provide food for hundreds of millions of people, a barrier of defense against the ravages of tropical cyclones and tsunamis, and a tremendous source of tourism income for nations lucky enough to have them grace their coastlines: the IPCC calculates that global coral-reef-associated tourism revenue is approximately $11.5 billion annually.

But all is not well with coral reefs, and scenes like the ones depicted on the following pages have become all too common. The IPCC concludes that "coral reefs are the most vulnerable marine ecosystem with little scope for adaptation." Studies conducted on reefs throughout the world document widespread reductions in coral coverage. The National Oceanic and Atmospheric Administration (NOAA) estimates that 20% of coral reefs are already damaged beyond recovery, and that 50% are in critical condition and under risk of collapse as ecosystems. Unless significant measures are taken to reduce the stress on coral reefs from human activities, global loss of coral reefs from most sites is very likely by the year 2050.

Healthy reefs
Coral reefs are among the most diverse and productive ecosystems on Earth. Unfortunately, healthy reefs such as the one shown here (from Indonesia) are becoming increasingly rare.

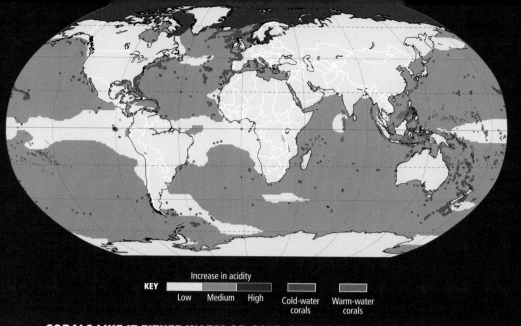

KEY

Increase in acidity

Low   Medium   High   Cold-water
corals   Warm-water
corals

CORALS LIKE IT EITHER WARM OR COLD, BUT NOT TOO HOT OR ACIDIC

Corals can be found in both warm and cold environments, but the highly diverse coral reefs like those pictured on the previous page are exclusively found in warm waters. Also shown is the effect of ocean acidification by the year 2100 (according to the high-emissions RCP 8.5 scenario), with the smallest increases in acidity in light blue and the largest in dark blue.

Unhealthy reefs

This coral reef is bleached: the corals have lost their symbiotic algae that give them their characteristic color. Scientists believe that coral bleaching is caused by excessively hot ocean temperatures.

(Cont.)

EFFECT OF DISEASE ON CORAL HEAD

Black band disease, a bacterial infection linked to warmer water temperatures, destroyed this massive coral head in the Florida Keys. The bacteria are concentrated in the black band at the boundary between healthy and dead coral, which migrates outward over time, and leaves dead coral behind.

1988

1998

Causes of decline

The causes of coral reef decline are many, and include:

Natural stressors

- Disease
- Predation
- Overgrowth by damaging algae

Human activities

- Overfishing
- Pollutant runoff from land
- Careless snorkelers walking on delicate coral
- Fuel spills
- Fuel and wastewater discharge from boats; oil spills

Additional natural factors are exacerbated by human activity. For example, coral bleaching—the loss of the algae that live in a symbiotic relationship with the coral animal and give it color—has been directly linked to intervals of exceptionally hot ocean temperatures. Human-induced global warming is likely contributing to this problem.

Marine protected areas

Marine protected areas (MPAs) are being established around the world, and have proven to be effective at staving off coral and fish losses. MPAs have been shown to be of great economic benefit as well. The United Nations Environmental Program estimates that MPAs cost less than U.S. $1000 per square km ($2500 per square mile), whereas the economic value of coral reefs has been estimated at U.S. $100,000–$600,000 per square km ($250,000–$1,550,000 per square mile).

Unfortunately, MPAs cannot protect coral reefs from warming, nor can they protect corals from the acidification effect of increases in atmospheric carbon dioxide. The ocean has absorbed approximately half of the CO_2 released by fossil-fuel burning and deforestation (◄ p.106). When CO_2 dissolves in water, it is transformed into carbonic acid, which makes the water less conducive to coral growth. Sadly, every square kilometer of the ocean, with the possible exception of a few remote

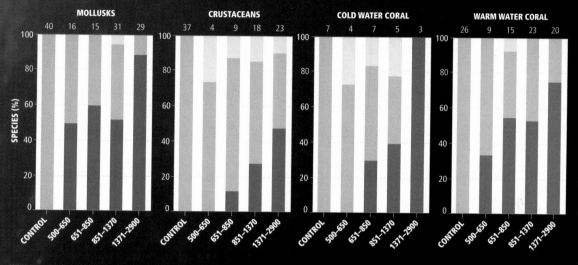

MOLLUSKS				
40	16	15	31	29

CRUSTACEANS				
37	4	9	18	23

COLD WATER CORAL				
7	4	7	5	3

WARM WATER CORAL				
26	9	15	23	20

SPECIES (%)

CONTROL 500–650 651–850 851–1370 1371–2900

KEY

- Positive effect
- No effect
- Negative effect

GROWTH RATE IN DECLINE

These bar charts show the sensitivity of various types of marine organisms to the buildup of atmospheric CO_2 by the year 2100. The numbers above the bars indicate the number of species studied. The control reflects no change from today. The other bars indicate the effects of progressively higher atmospheric CO_2 levels; 500–650 ppm (RCP 4.5; ◀ p.92), 651–850 ppm (RCP 6.0), 851–1370 ppm (RCP 8.5), and a very high CO_2 scenario. While some marine organisms might benefit from modest CO_2 increases, negative effects prevail at higher levels, 1371–2900 ppm.

Experimental studies suggest that if fossil-fuel burning rates continue to increase, corals will be unable to grow skeletons by the end of this century.

Arctic areas, has suffered detrimental consequences from human activities similar to those experienced by coral reefs.

The only way to prevent this catastrophe is to reduce or eliminate CO_2 emissions, or to develop effective means of sequestering CO_2 before it escapes into the atmosphere.

Coral or cars?

Humans face a stark choice: either we continue emitting carbon dioxide at ever-increasing rates and accept the

demise of coral reefs, or save the coral reefs by reducing or eliminating fossil-fuel CO_2 emissions. It's a genuine dilemma, but for those who value the beauty and understand the importance of coral reefs to the global environment and to society, the path is clear.

And it's not just coral reefs that are at stake. A common theme of this book is that for many reasons it may be in our own best interests to reduce the buildup of atmospheric CO_2.

The highway to extinction?

Gone for good?
The golden toad was last seen in the cloud forests of Costa Rica in 1989.

The diversity of species on planet Earth today is the result of millions of years of evolutionary interaction between life and its environment. Human intervention is a new, powerful force, which some liken to the forces that led to mass extinctions of life in the past, such as the asteroid impact that probably precipitated the demise of the dinosaurs 65 million years ago.

Polar bears in danger

A case in point is the precarious future of the polar bear, which depends on expansive sea-ice cover to reach and feed on seals. The earlier spring breakup and retreat of the sea-ice now forces polar bears to remain on the tundra, where they must fast and survive on reserves of fat. This puts particular stress on female polar bears, which spend the winter in nursing dens and need easy access to seals in spring to rebuild their fat reserves.

Temperature changes and limited water availability can stress individual organisms that find themselves suddenly outside of

Bears on ice
Since polar bears hunt on sea-ice, the melting of the Arctic ice cap is making it increasingly difficult for the bears to find sufficient food.

their climate "comfort zone." Typically it is not just one species that is affected. Nearly 60% of widespread and common plant species and 35% of widespread and common animal species will see their habitat ranges shrink by over 50% by 2080.

Extinct amphibians

Also of concern is the worldwide loss of amphibians. The golden toad was last seen in the cloud forests of Costa Rica in 1989. In cloud forest ecosystems, mist from clouds is the primary source of moisture. As the climate warms, trade winds rising up the mountain slope condense at higher elevations, so clouds shift upward. Cloudiness increases, but clouds no longer intersect the forest floor, so it gets drier but the nights get warmer. Birds, reptiles, and amphibians have all been affected, but the golden toad and many species of the harlequin frog are now believed to be extinct. Some scientists theorize that warmer nights may favor growth of the chytrid fungus—a potentially fatal pathogen that grows on amphibian skin.

Adapt or die

Amphibians are the first group of organisms identified as at risk of extinction from global warming. Many more will follow as the planet warms, especially if the rate of warming is rapid. Organisms adapt and ecosystems migrate at rates that sadly may be too slow to prevent ecosystem collapse and the extinction of species.

The IPCC 4th Assessment Report stated with medium confidence that 20–30% of plants and animals will be subject to increased risk of extinction if global temperatures rise to 2.0°C (3.6°F) above the pre-industrial level, and perhaps 40–70% of species will be at risk of extinction if temperatures rise by 4.0°C (7.2°F). The current report expresses high confidence in these conclusions.

We must remember that extinction is irreversible, and that we are inseparably dependent on the diversity of species harbored by our planet, and the goods and services provided by the ecosystems they support (◀ p.124).

SPEED LIMITS OF MIGRATION

The rates at which climate zones are expected to move under various fossil-fuel emissions scenarios show that climate speeds exceed the natural migration capacities of most plants and animals for the higher-emissions scenarios. Climate zone migration is faster across flat areas (areas with less mountainous relief).

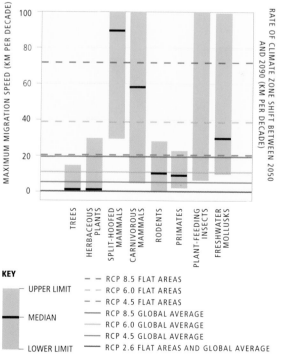

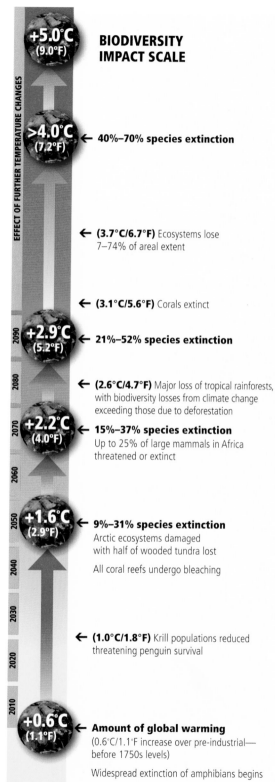

BIODIVERSITY IMPACT SCALE

+5.0°C (9.0°F)

>4.0°C (7.2°F) ← 40%–70% species extinction

← (3.7°C/6.7°F) Ecosystems lose 7–74% of areal extent

← (3.1°C/5.6°F) Corals extinct

+2.9°C (5.2°F) ← 21%–52% species extinction

← (2.6°C/4.7°F) Major loss of tropical rainforests, with biodiversity losses from climate change exceeding those due to deforestation

+2.2°C (4.0°F) ← 15%–37% species extinction
Up to 25% of large mammals in Africa threatened or extinct

+1.6°C (2.9°F) ← 9%–31% species extinction
Arctic ecosystems damaged with half of wooded tundra lost

All coral reefs undergo bleaching

← (1.0°C/1.8°F) Krill populations reduced threatening penguin survival

+0.6°C (1.1°F) ← **Amount of global warming**
(0.6°C/1.1°F increase over pre-industrial—before 1750s levels)

Widespread extinction of amphibians begins

Too much and too little
Will floods and droughts really get worse?

There is nothing more precious to living things than water. Changes in the availability of fresh water are of paramount importance in gauging the impacts of climate change on society. These impacts may at first seem contradictory, since increased drought is predicted in many regions, while more frequent intense precipitation events and flooding are predicted for others (◀ p.98). Such diverse changes result from a complex pattern of shifting rain belts, more vigorous cycling of water in a warmer atmosphere, and increasing evaporation from the surface due to warmer temperatures.

In July 2014, a toxic algae bloom in Lake Erie led to a ban on the use on use of tap water for drinking, cooking, or bathing for more than half a million residents of Toledo,

Sunken city
An aerial view shows a submerged trailer park after the Mississippi River flooded the city of Memphis, Tennessee, in April and May in 2011. Flooding is likely to become more commonplace in many parts of North America and Europe, even in places that are also suffering from drought.

One of the most significant potential impacts of climate change is diminished or unreliable fresh water supplies.

Shallow river

As a result of severe drought in the Brazilian Amazon, rivers that are a lifeline for local people for transport, supplies, and trading are drying up. Here, thousands of dead fish clog the edges the Manaquiri River, a tributary of the Amazon.

Ohio. Such events have become more frequent due, in part, to warming of fresh water bodies. Globally, water demand is likely to escalate significantly in future decades, primarily due to population growth. Yet this growth is taking place at a time when, in many regions, fresh water resources may be growing more scarce due to climate change.

A combination of warmer water, more intense rainfall events, and longer periods of low river levels and stream flows will also exacerbate water pollution. Combined with other aggravating factors, such as population growth and increased urbanization, these impacts put intense pressure on fresh water supplies. Since steady running water is required for hydroelectric energy plants, and for cooling towers used in nuclear energy production, decreased water flow can threaten energy resources, too.

» (Cont.)

»
(Cont.)

Serious negative impacts

The negative impacts of changing precipitation patterns outweigh the benefits. For example, the increases in annual rainfall and runoff in some regions are offset by the negative impacts of increased precipitation variability, including diminished water supply, decreased water quality, and greater flood risks. There is hope, however, that in some cases, adaptations (e.g., the expansion of reservoirs) may offset some of the negative impacts of shifting patterns of water availability (▶ p.160).

WORLDWIDE
More than 15% of the world's population depends on the seasonal melt of high elevation snow and ice for fresh water. The melting of glaciers and ice caps (◀p.110) represents a serious threat.

MORE FREQUENT EXTREME DROUGHT EVENTS

Climate model simulations predict that the spacing between consecutive extreme drought events (defined as once-in-a-hundred-years events) will decrease sharply in many regions by the 2070s for the "middle of the road" emissions scenario (◀p.92).

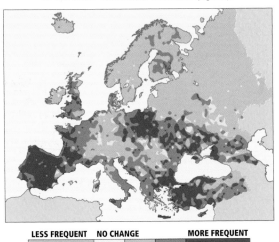

LESS FREQUENT	NO CHANGE			MORE FREQUENT
<	100	70	40	10

SIMULATED RETURN PERIOD (TYPICAL NUMBER OF YEARS BETWEEN CONSECUTIVE DROUGHTS) FOR EXTREME DROUGHT (I.E., DROUGHT WITH A MAGNITUDE EQUAL TO WHAT IS CURRENTLY CONSIDERED A 100-YEAR DROUGHT) BY LATE 21ST CENTURY (2070–2079).

FUTURE CLIMATE CHANGE IMPACTS ON WATER

U.S.
A steady increase in the population of cities in desert climates such as Phoenix and Las Vegas is occurring at precisely the same time that drought conditions are worsening.

SOUTH AMERICA
Aquifers will be depleted 75% by 2050.

SOUTHERN EUROPE AND THE MEDITERRANEAN

Many arid and semi-arid regions, such as the Mediterranean and parts of southern Europe, southern Africa, and much of Australia, are likely to suffer from increased drought. Electricity production potential at hydropower stations may decrease by more than 25% by 2070.

AFRICA

The spread of disease will increase due to more heavy precipitation events in areas with poor water supplies and an overtaxed sanitation infrastructure.

MEAN CHANGE OF ANNUAL RUNOFF, IN PERCENT, BETWEEN THE PRESENT (1981–2000) AND 2081–2100 (SIMULATED)

| < | -50 | -30 | -20 | -10 | -5 | 0 | 5 | 10 | 20 | 30 | 50 | > |

INDIA

In many coastal regions there will be plenty of water, but it will be the wrong kind! A sea level rise of 0.1 m (0.3 ft) by 2040–2080 will threaten the fresh water supply.

BANGLADESH

Areas with increased rainfall and runoff will suffer from an enhanced risk of flooding. The impacts are likely to be especially harsh for regions like Bangladesh, which is already facing the pressures of rising sea level (◀ p.110).

Is warming from carbon dioxide leading to more air pollution?

Carbon dioxide isn't a pollutant in the typical sense—that is, something introduced into the environment that is a direct threat to human health or to nature. The most notable detrimental health effects of rising CO_2 levels are indirect. They include the negative health effects related to warming, which we will discuss in more detail (▶ p.142), and the intensification of air pollution—a more recently discovered phenomenon.

Comprehensive models of climate, pollutant chemistry, and human health effects are being used to calculate the relationships between atmospheric warming and the buildup of pollutants. By directly calculating human health effects, these models are different from standard climate models.

Smog is produced when emissions from incomplete fossil-fuel combustion react to produce pollutants. One such pollutant of note is tropospheric ozone, a lung irritant that also damages crops, buildings, and forests. (Ozone in the stratosphere occurs naturally and protects us from ultraviolet radiation, but near the ground it acts as a pollutant.) Warming accelerates ozone production and promotes air stagnation, leading to increasing tropospheric ozone levels. One study mentioned in the IPCC report attributed 50% of the deaths in the European heat wave of 2003 to intensified ozone pollution rather than from the heat itself.

One comprehensive model indicates that for each 1°C (1.8°F) increase in

In the U.S., every 1°C (1.8°F) temperature rise will result in 1000 extra pollution-related deaths each year.

temperature, there will be an additional 20,000 pollution-related deaths worldwide annually. This same amount of warming results in even more notable increases in the incidence of asthma and other respiratory illnesses. In the model simulation, there was no question that the cause of these health problems was the buildup of CO_2, because that was the only change to which the model was subjected.

Subsequent studies have shown that aggressive CO_2-reduction strategies, like that reflected by RCP2.6, could save hundreds of thousands of premature deaths by ozone pollution annually. And a reduction in methane emissions by 20% would not only reduce warming, it would reduce ozone levels (because atmospheric chemical reactions involving methane produce ozone) and save 370,000 lives between now and 2030.

Los Angeles sunset
Although significant progress has been made to reduce smog in Los Angeles, greenhouse warming may jeopardize future efforts to clean up the air here and in other large polluted cities such as Houston and Mexico City.

War...

Some climate change skeptics believe that global warming is the sole domain of conservationists, tree-huggers, and utopian idealists. Yet policy experts in national security and global conflict, including former CIA directors and White House chiefs of staff, also worry about the potential threats posed by fossil-fuel burning.

Why? First, there is the most immediate and obvious reason: reliance on fossil fuels threatens the security of many developed nations by placing them at the mercy of often volatile regimes. And there is another, less-discussed security reason: an open Arctic Ocean, forecast to be the norm in "middle of the road" emissions scenarios (▶ p.148), would have clear international implications. If the forecasts are correct, North American and Eurasian nations will suddenly have new northern coastlines to defend.

But perhaps the most important security concern is the potential for increased competition among nations for diminishing essential resources. As any student of history can tell you, in many parts of the world, increased stress on resources has historically led to local unrest and unstable regimes, such as in Latin America.

The possibilities for conflict are countless. For example, the predicted changes in precipitation patterns will naturally create increased competition for available fresh water. Imagine the current Middle East political strife with the added facet of vicious water-resource competition.

OPENING THE NORTHWEST PASSAGE
As Artic sea-ice retreats and the once-fabled "Northwest Passage" opens up, new sea routes connecting the north Pacific and north Atlantic Oceans are becoming available.

Environmental refugees
A future with expanded patterns of drought and conditions unfavorable for agriculture and farming is likely to change people's notions of what constitutes desirable land. Sea level rise and other factors will also potentially make currently inhabited regions inhospitable to humans, thereby increasing competition for habitable land.

The term "environmental refugee" was coined to describe individuals fleeing their homelands for more benevolent conditions and climes. Lest one think this merely a hypothetical concept, it should be noted that an estimated 25 million environmental refugees have already been displaced. This

Diomede Islands
The Diomede Islands sit almost in the center of the Bering Strait—the gateway between the Arctic and Pacific oceans. The strait is likely to become a busy shipping route as the Northwest Passage opens up.

is more people than have fled civil war or religious persecution in recent years. Climate change appears to be driving the ongoing migration from the dry Sahel to neighboring regions of West Africa. It also appears to be playing a role in the exodus of people from parts of India, China, Central America, and South Africa. The aftermath of Hurricane Katrina serves as a reminder that even the industrialized world may not be immune to climate refugeeism.

Ripe for conflict

An optimist might hope that the global threat of climate change will unite the international community as never before, spurring a coordinated campaign among nations to save humanity. Unfortunately, conflict experts foresee the possibility of a different scenario.

A global population predicted to increase to about 9 billion by the mid-21st century, combined with stresses on water, land, and food resources could create the "perfect storm."

As nations around the world exceed their capacity to adapt to climate change, violence and societal destabilization could ensue, leading to unprecedented levels of conflict both between and within nations.

A combination of worsened drought, oppressively hot tropical temperatures, and rising sea levels could displace a large enough number of people by the mid-21st century to challenge the ability of surrounding nations to absorb them, with political and economic turmoil ensuing.

One possible scenario, for example, is that increasingly severe drought in West Africa will generate a mass migration from the highly populous interior of Nigeria to its coastal mega-city, Lagos. Already threatened by rising sea levels, Lagos will be unable to accommodate this massive influx of people. Squabbling over the dwindling oil reserves in the Niger River Delta combined with potential for state corruption will add to the factors contributing to massive social unrest.

Another possible scenario is that drought and decreased river runoff in southwestern North America will strain already water-poor and resource-starved Mexico, leading to increased migration to the U.S. and stress on already delicate diplomatic relations between the two countries.

Even more ominous conflicts can be imagined for the more extreme climate change scenarios. As the nations and peoples of the world compete for diminishing resources, it may become increasingly difficult to establish or maintain stable governance. Indeed, some experts have described worst-case scenarios not so different than those in post-apocalyptic fables such as *Mad Max* and *Oryx and Crake*.

Famine...
More people, less water, less food

Climate change has the potential to seriously undermine the world's food supplies. Sadly, many of the regions most likely to be affected already find it difficult to meet existing food demands.

Short-term positives, long-term negatives

Perhaps surprisingly, some regions, such as the U.S., Canada, and large parts of Europe, stand to benefit at moderate levels of additional warming (1.0–3.0°C/1.8–5.4°F), thanks to increased crop and livestock productivity.

For these lucky countries, warming will result in longer growing seasons, favorable shifts in rainfall patterns, and higher CO_2 levels, which provide a short-term benefit for plant growth. Indeed, global food production is projected to increase on average with modest levels (1.0–2.0°C/ 1.8–3.6°F) of future warming. But at even those moderate levels, many tropical and subtropical regions—including India and sub-Saharan and tropical Africa, which already struggle to meet food and pasturing demands—will likely experience a combination of warmer temperatures and decreased rainfall. This will cause a corresponding decrease in productivity of key crops, such as the major cereals.

PROJECTED CLIMATE CHANGE IMPACTS ON CROP AND LIVESTOCK YIELDS

Expected changes in crop and livestock yields by 2050 given mid-range emissions. Note that tropical regions suffer more losses than temperate regions due to climate change.

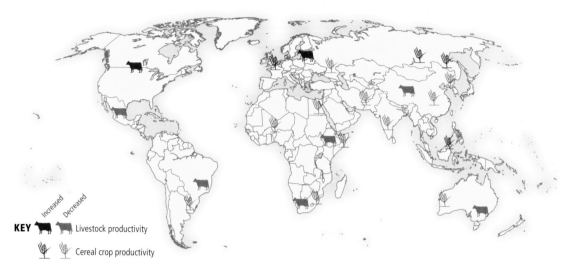

KEY · Livestock productivity

· Cereal crop productivity

Increased · Decreased

Moreover, the substantial warming (more than 3.0°C/5.4°F) projected to beset us as early as the late 21st century in the "business-as-usual" emissions scenario (◀ p.92) will probably have negative impacts for food crops in all major agricultural regions. Increased forest fires (▶ p.144) and more frequent disease and pest outbreaks (▶ p.142) could also diminish available food resources. What's more, the agricultural labor supply is likely to be disrupted by population displacement due to flooding (◀ p.110) and epidemics.

At sea

Fish populations, including commercial and subsistence fishing, may be greatly affected by climate change. The impact of warming is likely to be variable, leading to either increases or decreases in aquatic populations, depending on the location and the species of fish present. Changing ocean circulation patterns also represent a wild card. In particular, the potential weakening in the "conveyor belt" pattern of ocean circulation in the North Atlantic could have significant negative consequences for fisheries in that region.

Ray of hope

Socioeconomic development could partially or even completely offset the negative impacts of climate change on the food supply. There are roughly 820 million people around the globe who are currently undernourished. Taking into account the combined impacts of changing socioeconomic factors and climate change, projections for the number of undernourished people by 2080 range from a reduction to 100 million to an increase to 1.3 billion.

Somalian famine victims
The line for food in Baidoa, Somalia seems endless. Unfortunately, climate change will only make situations like the devastating famine in Somalia worse.

...Pestilence and death
Human health and infectious disease during times of global climate change

We install heating and air conditioning for indoor climate regulation, build dams for flood control, dig wells to irrigate our fields, and construct dikes to stave off rising seas. But despite our best efforts to insulate ourselves from our natural environment, we remain highly susceptible to climate change, especially when its effects are rapid or unexpected.

Pests and pollen
Of paramount concern is the effect that climate change might have on human health. Diseases can spread as climates change—insects and rodents that carry disease range more widely as climate barriers are lifted. Already there is evidence that vectors such as ticks are spreading to higher latitudes and altitudes in Canada and Sweden. And ragweed is producing more pollen over a longer season as a result of rising temperatures and atmospheric CO_2 concentrations.

Disease-carrying mosquitoes typically confined to warmer climes are spreading into the extratropics due to warmer winters, carrying infectious diseases such as dengue and West Nile virus. The first recorded West Nile virus outbreak in the U.S. was in the summer of 1999, following the record warm 1998–99 winter. In 2012, the warmest year on record to date for the U.S., the largest number of West Nile–related deaths (286) on record occurred. Other vector-born diseases, such as malaria, could expand their range within the tropics as warmer temperatures lead to a faster parasite reproduction cycle.

Heat can kill
The wake-up call for climate-change-induced human mortality is the European heat wave of 2003 (◄ p.58). During two extremely hot weeks in early August 2003, nearly 15,000

Some like it hot
Mosquitoes and other vectors of disease may spread and flourish as global temperatures increase.

EFFECTS ON HUMAN HEALTH

The IPCC report projects the following climate changes and related health effects in the 21st century.

Predicted climate change (in order of decreasing certainty)	Anticipated effect on human health
On land, fewer cold days and nights	Reduced mortality from cold exposure (virtually certain)
More frequent heat waves	Increased mortality from heat, especially among the young, elderly, infirm, and those in remote regions (virtually certain)
More frequent floods	Increased deaths, injuries, and skin and respiratory disease incidence (very likely)
More frequent droughts	Food and water shortages; increased incidence of food- and water-borne disease and malnutrition (likely)
More frequent strong tropical cyclones	Increased mortality and injury, risk of food- and water-borne disease, and incidence of post-traumatic stress disorder (likely)
More extreme high-sea-level events	Increased death and injury during floods; health repercussions of inland migration (likely)

French people died; across Europe, fatalities approached 35,000. Most of the dead were elderly people who were unable to escape the persistent and oppressive heat. The death rate began climbing several days after temperatures began to rise, and peaked at over 300 additional deaths per day in Paris alone before temperatures finally began to fall. During the summer of 2010, an anomalous, large-scale weather pattern resulted in over 15,000 deaths in Russia due to a heat wave, while thousands more were killed by record flooding in Asia. The IPCC report projects an increase in heat-wave incidence with high confidence.

Climate-change-related health impacts will not be uniformly distributed across the world's population. Poor nations will be more susceptible than wealthy ones because of inadequate access to air conditioning, infrastructure (clean water supplies, electricity, etc.), health care, and emergency response facilities. In all countries, children, the elderly, and the urban poor will suffer disproportionately, as will those people living in low-lying coastal areas.

These threats to human health may serve as a motivating factor for governments to mitigate future climate change. Potential adaptations include raising public awareness, instituting advance warning systems, and improving public health infrastructure in those regions most likely to be hardest hit.

Earth, wind, and fire

Impacts on North America

In North America, annual costs associated with weather-related damage run to tens of billions of dollars. The predicted increase in extreme weather events (◀ p.112), combined with sea level rise and other climate change impacts, represents a serious threat to people, infrastructure, ecosystems, and the economy.

PATTERN OF TEMPERATURE CHANGE OVER NORTH AMERICA IN RECENT DECADES

The greatest warming is found in the northwest, but all regions have warmed.

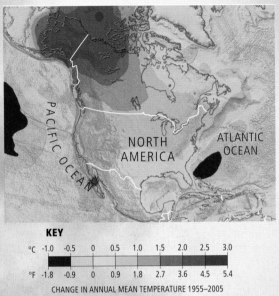

KEY

°C	-1.0	-0.5	0	0.5	1.0	1.5	2.0	2.5	3.0
°F	-1.8	-0.9	0	0.9	1.8	2.7	3.6	4.5	5.4

CHANGE IN ANNUAL MEAN TEMPERATURE 1955–2005

TRENDS IN NORTH AMERICA IMPACTED BY RISING TEMPERATURES

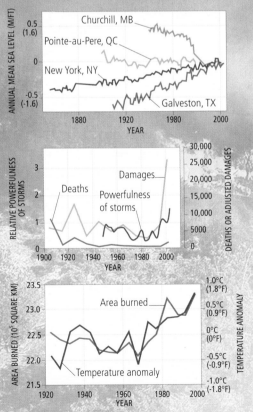

Earth

This graph shows how relative sea levels on the North American coast have changed over the last century. In some regions, such as eastern Canada, the rising of the coastline due to Earth's slow rebound from the last ice age has largely offset the sea level rise resulting from global warming. In other regions, such as the Gulf Coast of the United States, this rebound is having the opposite effect and compounding sea level rise.

Wind

Hurricane energy and powerfulness have increased in recent times in the United States. It is interesting to note that while damage from storms is increasing—not only due to escalating storm energy and power, but also because of growing coastal populations and more coastal development—mortality hasn't. This is thanks largely to better warning systems and evacuation measures. Still, punishing storms have the potential to wreak havoc, especially on property and infrastructure, in many coastal communities.

Fire

This graph compares trends in surface temperature and forest area burned in Canada over the past 100 years. Warming temperatures mean longer fire seasons, larger forest fires, and a heightened threat to human communities and forest ecosystems. A confluence of record-breaking heat, drought, and fuel load has led to unprecedented wildfires in the western U.S. in recent years. Extreme heat and dryness in Eurasia led to an unprecedented outbreak of hundreds of wildfires in Russia during the summer of 2012. The combined record heat and smoke generated smog that led to over 50,000 deaths across Russia. Damages were estimated to exceed U.S. $15 billion.

Earth

The progressive inundation, storm-surge flooding, and shoreline erosion associated with rising sea level are increased threats to coastal regions. The impacts will be particularly significant along the Gulf of Mexico, and the southern and mid-Atlantic coastlines of North America. In these areas, local sea level is already rising due to Earth's slow rebound from the ice-weight load of the last ice age. (In the extreme northeast, this rebound is acting in the opposite direction, lowering local sea levels.) Rapidly growing coastal populations and overburdened infrastructure make these regions highly vulnerable to the impacts of climate change on property and life.

Wind

Tropical cyclones are likely to be more powerful along the Gulf and Atlantic coasts (◀ p.62), where the coastal ecosystems, populations, and infrastructure are already threatened by sea level rise.

Fire

Wildfires in North America have become more common in recent decades, and their incidence is likely to rise still further in response to climate change impacts such as increased drought, decreased winter snow-pack (the amount of winter snow that accumulates at higher elevations), earlier melt (leading to a longer fire season), and increased fuel load from forests killed or weakened by pests such as the Pine Bark Beetle that have expanded their range due to warming winters. For mid-range climate change scenarios, it is predicted that the forest burning area in Canada could roughly double by 2100. Burgeoning populations in regions near the boundary between urban and forested areas will increase human vulnerability to wildfire.

The wider world

Australia and New Zealand may suffer similar impacts to North America, such as intensified droughts and sea level rise. Decreased soil moisture due to warming temperatures will likely lead to increases in the frequency and severity of wildfire in those regions, with negative impacts on agriculture and forestry. As in North America, the significant expansion of infrastructure and population in coastal regions (e.g., Cairns and southeast Queensland in Australia, and the coastline from Northland to the Bay of Plenty in New Zealand) will increase society's vulnerability to rising sea level and (potentially more powerful) tropical cyclones in future decades.

Africa, Europe, and Asia, and some island nations also have coastal populations that are threatened by sea level rise and more frequent and powerful tropical cyclones. Drier summers over Africa, large parts of Europe and the Mediterranean, and the Amazon basin may lead to increased fire risk as well.

Colorado on fire
The large East Peak fire rages near La Veta, Colorado, in 2013. The total area affected by forest fires in the western U.S. has increased by more than a factor of six in the past two decades.

Too wet and too hot
Impacts on Europe

The impacts of climate change are already apparent across Europe, from the record-setting heat wave of summer 2003 (◀ p.58) and the melting of long-standing mountain glaciers, to shifting precipitation patterns and observable changes in ecosystems. Some of these changes can be attributed to an atmospheric phenomenon known as the **North Atlantic Oscillation (NAO)**. The increasing tendency for the North Atlantic Oscillation to be in its positive phase has favored a stronger, more northerly jet stream over Europe. This may, at least to some extent, be associated with climate change.

Too wet

Average annual rainfall has already increased by nearly 50% over parts of northern Europe. Further increases in winter flooding in coastal regions, and an increased frequency and intensity of flash floods for much of Europe are projected with further warming. Almost two million people in the low-lying countries of the Netherlands, Belgium, and Luxembourg could be threatened with flooding in the coming decades as a result of these changes combined with the impacts of rising sea levels.

SELECTED POTENTIAL CLIMATE CHANGE IMPACTS IN EUROPE

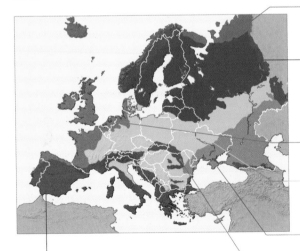

- Thawing of permafrost
- Substantial loss of tundra biome
- More coastal erosion and flooding

- More coastal flooding and erosion
- Greater winter storm risk
- Shorter ski season

- More coastal erosion and flooding
- Stressing of marine ecosystems and habitat loss
- Increased tourism pressure on coasts
- Greater winter storm risk and vulnerability of transport systems to high wind conditions

- Increased frequency and magnitude of winter floods
- Heightened health threat from heat waves

- Decreased crop yield
- More soil erosion
- Increased salinity of inland seas

- Severe fires in drained peatland
- Disappearance of glaciers
- Shorter snow-cover period
- Upward shift of tree line
- Severe biodiversity losses
- Shorter ski season
- More frequent rock slides

- More frequent forest fires
- Biodiversity losses escalate
- Negative impact on summer tourism
- Heat wave impacts grow more serious
- Cropland losses as well as losses of lands in estuaries and deltas

The average rainfall has already increased by 50% over parts of northern Europe.

Too hot

For Europe, the past decade has been the warmest on record for at least the last 500 years. Climate model projections indicate that further warming will be concentrated in northern Europe in winter and in southern and central Europe in summer.

It is possible that certain impacts of these changes could be beneficial for human inhabitants. For example, warmer winters could reduce the number of deaths arising from exposure to extreme cold. On balance, however, it appears probable that the risks to human health will increase. Deadly heat-stress, associated with events such as the 2003 heat wave, will almost certainly become more common in Europe. Increased flooding and warmer winters are also likely to facilitate the spread of water-born, vector-born, and food-borne diseases (◀ p.142).

Across the globe

Already common, heat waves are likely to increase in frequency and intensity over the next few decades as large parts of North America, Asia, Australia, and Africa continue to warm. In the world's cities, urban "heat island" effects, poor air quality, population growth, and an aging population are likely to magnify the impacts of rising temperatures.

Over large parts of North America, Asia, Australia, New Zealand, and South America (e.g., Venezuela and Argentina), increases in heavy rainfall have already led to a higher incidence of floods and landslides. Further increases will likely lead to degraded water quality in these regions (◀ p.132) and the spread of water-borne diseases.

Wettest winter on record
Hastily built flood defenses surround a house on the Somerset Levels in the UK. In the UK, where records go back two-and-a-half centuries, the winter of 2013/2014 was the wettest on record. Scientists from the UK Met Office have stated that climate change almost certainly played a role.

The polar meltdown

Every decade for the last 50 years, the snow-free period in Arctic Eurasia and North America increases by five or six days, exposing dark ground that absorbs sunlight and warms the soil. At the same time, the Arctic Ocean's summer minimum sea-ice cover decreases at 9–14% per decade, exposing the darker sea surface so that the Arctic warms even more. These and other positive feedback loops (◀ p.24) amplify the polar response to the buildup of fossil-fuel CO_2 (◀ p.102). On average, the Arctic is warming at twice the rate of the globe as a whole, and the region's vast land masses are warming at five times the global average. The rate of loss of the West Antarctic and Greenland ice sheets has increased markedly over the last two decades of observation.

Permafrost thaw

Below the Arctic's upper layer of soil, which thaws each year, is permafrost—permanently frozen soil. Permafrost is important to human settlement in the Arctic. It has provided a solid and impervious substrate to support building foundations, roadbeds, and pipelines, and to contain sewage ponds and landfill leachate (water that collects contaminants as it trickles through waste).

As a result of recent climate warming and locally released heat from buildings and pipelines, permafrost is now melting.

Thawing leads to building collapse, pipeline breakage, roadway degradation, and contamination of surrounding environments. Moreover, methane (CH_4)—a strong greenhouse gas trapped within the permafrost for millennia—is now being released into the atmosphere, increasing warming. By the end of the 21st century there will be a reduction of between 37% (RCP 2.6) and 81% (RCP 8.5) in the northern hemisphere permafrost area, releasing up to 250 billion **metric tons** of carbon as CH_4 or CO_2.

Meltdown pros and cons

The meltdown may benefit society with lower heating costs, greater opportunities for agriculture and forestry, increased river flows (supporting hydroelectric power generation and improved navigation through the Arctic), and easier extraction and transportation of marine resources (including potentially huge hydrocarbon reserves below the Arctic Ocean). But sea-ice retreat compounded by the rising sea level will also lead to coastal erosion and collapse of roads, buildings, electricity grids, and pipelines. Facilities, and even whole villages and towns, may need to relocate. Traditionally, indigenous Arctic people lived in small, widespread communities that could easily relocate when necesary. Modern settlement trends are different: two-thirds of the Arctic population lives in settlements of more than 5,000 inhabitants. This settlement pattern, combined with dependency on infrastructure, reduces the resilience of Arctic populations to environmental change.

Melting away

Coastal erosion is a growing problem in the Arctic. This permafrost coastline near Kaktovik, on the north shore of Barter Island, Alaska, clearly shows the effects of erosion on the soft soil.

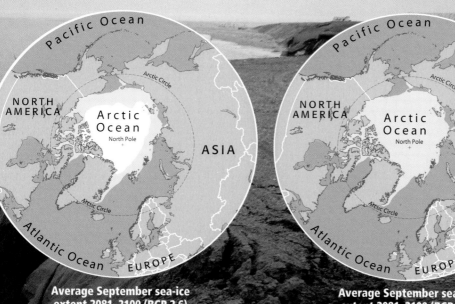

Average September sea-ice extent 2081–2100 (RCP 2.6)

Average September sea-ice extent 2081–2100 (RCP 8.5)

KEY

Average minimum 1986–2005

Average minimum 2081–2100

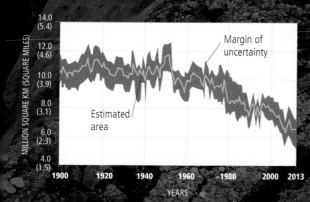

Margin of uncertainty

Estimated area

MILLION SQUARE KM (SQUARE MILES)

14.0 (5.4)
12.0 (4.6)
10.0 (3.9)
8.0 (3.1)
6.0 (2.3)
4.0 (1.5)

1900   1920   1940   1960   1980   2000   2013

YEARS

Observations of summer sea-ice cover

SEA-ICE COVER

The graph at left is based on observations from ships, aerial reconnaissance, land-based proxy information, and, since 1979, satellites. They reveal that the summer average sea-ice cover in the Arctic was fairly stable through the first half of the 20th century but then declined for decades, falling abruptly in 2007, causing global concern about the future of the Arctic. Models didn't project this level to be reached for several decades to come. The following year saw some recovery, but the sea-ice cover decline resumed. In the maps above, models project that the Arctic Ocean will be ice-free in September later this century for RCP 8.5; the smallest projected decline is 43% in year-round sea-ice cover for RCP 2.6. The pale blue area shows the 1986–2005 average minimum sea-ice extent for comparison.

Polar politics

The Arctic has gotten political attention in recent decades; large, untapped reserves of fossil fuels, minerals, and diamonds await better access for exploration and exploitation. The Russian Federation sent a submersible to the depths of the Arctic Ocean to plant its flag, and Denmark asserts that the geologic feature that traps petroleum in the Arctic Ocean is an extension of their country. On a more altruistic note, the developed world is also beginning to recognize and express concerns that the pollution it produces is resulting in harmful health and environmental impacts in the Arctic.

Operation IceBridge

http://goo.gl/Ci9tQl

Part 4
Vulnerability and Adaptation to Climate Change

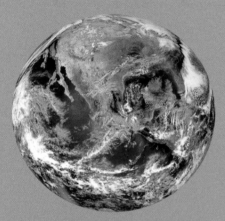

To reduce our vulnerability to climate change, it is necessary that we both adapt to the effects of the existing buildup of atmospheric CO_2 and reduce the amount of CO_2 we are emitting. In the initial stages of climate change, adaptations can help to allay the threat of rising sea level, diminished and shifting fresh water resources, loss of agricultural productivity, and adverse economic effects. Efforts to reduce or mitigate emissions face many obstacles, but ultimately they may be our best hope for the future, when adaptation alone will be insufficient to counter the long-term impacts of climate change on society, the environment, and the economy.

Key ideas

Adaptation

- Some economies can adapt to climate change more readily than others. China and Africa are currently the least adaptable and so most at risk, while most developed nations are better able to adapt.

- Sea-level rise is one of the biggest challenges posed by climate change. Even if emissions can be stabilized, many coastal and island communities will likely have to adapt by, for example, improving their sea defences.

- Increased global demand for fresh water coupled in some areas with reduced supply due to climate change will make it necessary to adapt water management practices.

- Meeting the food demands of a growing population while also adapting to a changing climate will require changes to the way food is grown, such as using new crop varieties and modifying planting locations and schedules.

Reducing emissions

- Mitigating against climate change will involve cutting CO_2 emissions in order to stabilize the amount of CO_2 in the atmosphere.

- Cutting emissions can have economic benefits, because the cost of reducing emissions might turn out to be less than the cost of dealing with the damage resulting from climate change.

Is global warming the last straw for vulnerable ecosystems?

How human activity has changed the rules of the game

Temperatures have been fluctuating on Earth since the planet developed an atmosphere. So why is climate change such a big problem all of a sudden? We know that the increasing ocean temperatures and atmospheric CO_2 levels linked to coral stress and mortality (◀ p.126) have geological precedence. Carbon dioxide levels have been fluctuating by 100 ppm for the last million years, and climates have changed as a result, yet corals have survived. Surely, given this past record of adaptation and survival, corals and other organisms will adapt to global warming?

Not necessarily. This "historical precedence" argument has several weaknesses, not the least of which is the fact that fossil-fuel carbon reserves have the capacity to boost atmospheric CO_2

levels higher and faster than in the last million years. There also are key differences between the modern adaptation "playing field" and that of the geologic past. Human land use, construction, pollution, and other constraints make adaptation and survival a different game than it used to be. Arctic ecosystems and coral reefs are particularly vulnerable. For example, coral reefs are under considerable additional stress from human activity such as:

- Wastewater and sediment discharge
- Pesticides
- Ship groundings
- Dynamiting and poisoning for fish collection
- Increasing recreational use

Is this any way to treat a reef?
This wastewater outflow pipe was discharging treated wastewater into the south Florida coastal zone. High nutrient levels in the discharged wastewater promoted algal growth that can smother coral reefs. Now these wastes are injected deep beneath the surface.

- In the Caribbean, a possible increase in pathogens carried by dust storms that have been exacerbated by human-caused desertification of Africa's Sahel region

These modern factors have increased the sensitivity of ecosystems to global warming and ocean acidification, and compromised their ability to bounce back from adversity.

In assessing the vulnerability of ecosystems to climate change we must consider:

- The potential rate of change
- Additional stresses imposed by human activity
- Barriers to adaptation and migration imposed by human activity, human settlement, and infrastructure (e.g., roads and pipelines)

Ecosystems in jeopardy

Will we act to reduce the stresses on ecosystems in advance of significant impact? Will we attempt to restore destroyed habitats or create new habitats before changes are irreversible? Are such efforts likely to be successful? There is much that we do not understand about how ecosystems function and what services they provide, despite considerable study (◄ p.124). Ecosystems have some capacity to adapt, and there is already evidence that many species are on the move toward higher altitudes or higher latitudes. But rates of adaptation and the ultimate capacity of organisms and ecosystems to adapt to rapid environmental change are limited (◄ p.130). Many think that if we don't act now, the anticipated stress of climate change will be "the straw that broke the camel's back."

What is the best course for the coming century?

Up to 1 m (3.3 ft) of sea level rise could take place by 2100, depending on which future emissions scenarios one considers (◀ p.98), and levels will continue to rise for many decades thereafter.

To adapt or to mitigate?

It seems that we have two options: adapt to these changes or mitigate against them. We can take actions to reduce the buildup of carbon dioxide in the atmosphere (mitigation) or to offset the effects of this buildup (adaptation). The best approach is to adopt a plan for the future that is a blend of both adaptation and mitigation, accompanied by technological development to support these efforts and research to guide the technology.

The IPCC report describes the vulnerabilities of countries to climate change with and without mitigation efforts, and assuming current or enhanced adaptive strategies (see maps right).

Adapt

The vulnerability of each system (that is, each country or region) is related to its adaptive capacity—its "ability or potential to respond successfully to climate variability and change." China and Africa currently have higher vulnerabilities than developed nations and will continue to have those vulnerabilities for the next few decades

Assuming current adaptive capacities and no mitigation efforts

VULNERABILITY LEVEL IN 2100

Vulnerability to climate change can be lessened if mitigation efforts are made and adaptive capacities are enhanced.

KEY (vulnerability level)

10 Extreme	7 Moderate	4 Modest
9 Severe	6 Moderate	3 Little
8 Serious	5 Modest	No data

With enhanced adaptive capacity and no mitigation efforts

because of their low adaptive capacity. Even though the adaptive capacities of developing nations are expected to improve with time, overall vulnerabilities are still predicted to remain high. In fact by 2100, even the nations in the developed world—including the U.S.—could be overwhelmed unless steps are taken to enhance adaptive capabilities and to mitigate against the buildup of CO_2 in the atmosphere.

The road ahead
The nations of the world are making decisions now that will make all the difference in how severe the consequences of climate change will be.

Mitigate

In the example illustrated by the maps below, mitigation means a stabilization of atmospheric CO_2 at 550 ppm (◀ p.116). This can realistically be achieved only by significant reductions in fossil-fuel burning rates. Decisive action to mitigate against CO_2 buildup leads to a substantial reduction in vulnerability in all but select regions of Africa, China, and Europe. Over the next few decades, mitigation efforts will primarily benefit developing countries. However, by the end of the century, all nations will benefit from taking such actions, as well as from investments to enhance their adaptive capacity.

Global action required

To improve their adaptive capacity, nations need to incorporate the likely effects of climate change into their strategies for sustainable development and disaster management. Mitigation efforts involve global factors such as atmospheric CO_2, so those efforts will require international agreements (▶ p.200). The IPCC report makes it clear that the nations of the world need to take aggressive and immediate action to avoid the looming crisis reflected in the maps shown here. Unfortunately, to date, some of the most resource-rich but climate-change vulnerable countries in the developing world have not committed to either mitigation or adaptation.

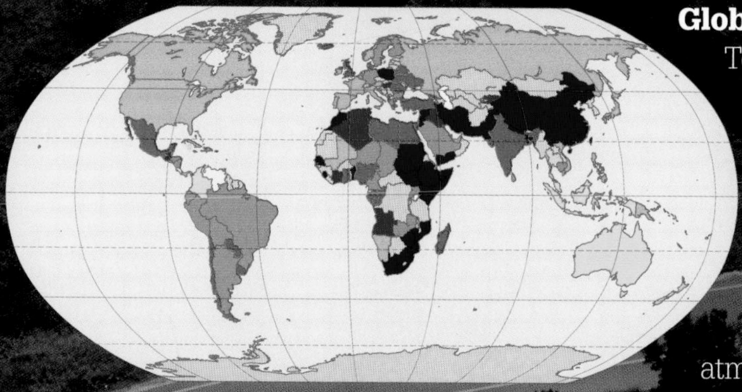

Assuming current adaptive capacities and mitigation efforts taken to stabilize atmosphere CO_2 levels at 550 ppm

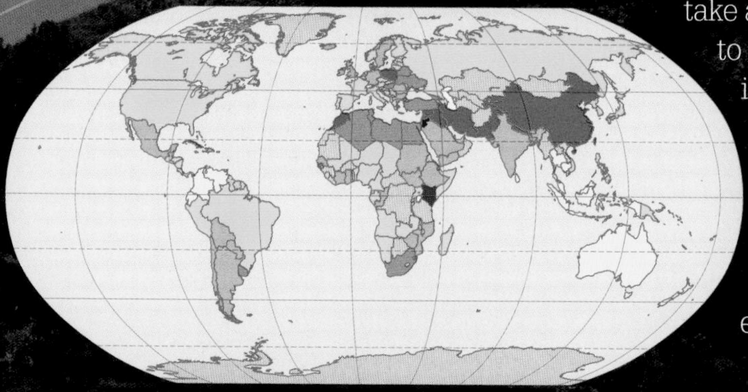

With both enhanced adaptive capacity and mitigation efforts taken to stabilize atmosphere CO_2 levels at 550 ppm

It's all about the economy!

It's clear that we can reduce the potential damage to natural ecosystems by reducing greenhouse gas emissions. What is less apparent is that emission reduction can be good for the economy as well. Economists tell us that the formidable cost of emissions reductions may actually be less than the economic damage that will result from climate change. Recent estimates of the economic damages from unbridled carbon emissions are 1–5% of GDP by the year 2100. According to the IPCC report, avoiding many of these damages by meeting the stringent target of 3.0°C (5.4°F) warming above pre-industrial temperatures (2.0°C/3.6°F further warming from 2014 temperatures) by 2100 would slow annual economic growth by only 0.06% (reducing growth from an estimated 2.3% to 2.24% per year).

The cost of carbon

A so-called "integrated assessment model," which takes into account economic considerations as well as climate change, can be used to estimate the "social cost of carbon," or SCC. This is the cost to society of emitting one additional metric ton (tonne)—1.1 U.S. tons—of carbon. The SCC incorporates the climate change and associated economic impacts of carbon emissions over a prescribed time horizon. This time horizon is typically up to the year 2100, but sometimes it also covers the centuries or millennia over which carbon emissions will continue to affect climate. Estimates of the SCC range from a few U.S. dollars per metric ton to several hundred. Current estimates of the SCC center around U.S. $30–$40. In other words, the typical American who drives 16,000 km (10,000 miles) per year and thereby emits a metric ton (1.1 U.S. tons) of carbon into the atmosphere imposes a cost of U.S. $30 on society. To add insult to injury, this cost impacts society both now and in the future—and the driver is not penalized at all for these damages. Thirty dollars may not seem like much, but think of all the drivers in North America alone, and it really adds up. And $30 may be an underestimate, because most calculations of the SCC do not consider the increased likelihood of extreme events like major floods, hurricanes, or abrupt sea-level rise and their associated costs.

Driving up carbon
The typical American drives about 16,000 km (10,000 miles a year), emitting about a metric ton (1.1 U.S tons) of carbon into the atmosphere as a result. The financial and environmental costs of this carbon are currently borne by society as a whole, rather than drivers.

Carbon credits

Emissions reduction passes a cost-benefit analysis only when the SCC exceeds the costs of carbon reduction—in the example above, that means as long as it costs less than U.S. $30–40 to offset the emissions of the typical American car driving 16,000 km

(10,000 miles) per year. Estimates of the immediate costs of mitigation fall below this threshold, but rise rapidly in the future, especially if we delay taking action to reduce fossil-fuel emissions. The SCC can be used to set the value of **carbon credits**—credits issued to nations for reducing carbon emissions, for example as part of the Kyoto Protocol—or level of taxation. If the SCC is U.S. $30, a 9 cents per gallon gasoline tax would offset the cost incurred by society for the damage done by driving 16,000 km (10,000 miles) a year. Prominent economist William Nordhaus equates such a tax with other taxes on harmful practices, such as smoking, and contrasts it to taxes on beneficial activities, such as labor.

Model limitations

Most of the integrated assessment models do not take into account the distinct possibility of abrupt climate change, which could lead to catastrophic damages without historical precedence. They also do not take into account the possibility that climate change will far exceed current projections. The resulting damages could lead to astronomical increases in the SCC.

The World Bank estimates that the cost of adaptation to climate change in the year 2050 will be U.S. $70–100 billion annually.

potential climate change damage over the next several decades
$20,000,000,000

−

the potential reduction
$5,000,000,000

+

the cost of reduction
$2,000,000,000

=

net climate change damage
$17,000,000,000

Ethical concerns

Finally, it should be noted that ethical concerns may call for action even when cost-benefit analyses do not. While many are projected to suffer some income loss as a result of climate change, the poor in developing nations are likley to experience far greater losses. The fact that climate change will probably redistribute resources (◀p.140) in a "reverse Robin Hood" fashion is particularly unfair to developing nations, such as Bangladesh. The cost of inaction to these communities may be incalculable.

Given this inherently unbalanced scenario, is it fair for the industrial nations—the primary generators of greenhouse gas emissions—to be the ones calling the shots and determining whether or not action is worth taking?

A finger in the dike

Up to 1.0 m (3.3 ft) of sea level rise could take place by 2100, given mid-range future emissions scenarios (◄p.110). And sea level might ultimately rise as much as 5.0–10.0 m (16.4–33.0 ft) if significant parts of the Greenland and West Antarctic ice sheets melt. Even though that could take several centuries to happen, we may be committed to this eventuality by 2100 if emissions meet or exceed the mid-range scenario estimates. Such a substantial rise in sea level would threaten the viability of coastal settlements worldwide (◄p.122).

Even a modest future sea level rise would be problematic, given the the current trend toward the aggressive development of coastlines. This impending crisis remains one of the biggest challenges posed by climate change. Adoption of the most austere stabilization policies could reduce the risks of higher-end projections, but we are already likely committed to moderate (that is, more than 0.3 m/1 ft) inundation. Some degree of adaptation will be required in addition to mitigation (▶ part 5).

Extreme weather

In December 2014 severe storms hit the northwest UK, with strong winds exceeding 160 kph (100 mph) and huge waves battering coastal areas, such as Prestwick, Scotland (as seen here). Faced with rising sea levels and the possibility of extreme weather events becoming more common, protecting at-risk coastal communities poses an increasing challenge.

PROTECTING COMMUNITIES FROM RISING WATER

There are three stages of adaptation that coastal communities threatened with rising sea levels may take.

Protection through engineering
The first stage—the most proactive —seeks to protect the population and infrastructure through engineering solutions (like the construction of "empolderings" that structurally reclaim inundated land, and coastal defenses such as dikes or beach nourishments, which create impediments to inundation).

Accommodating inundation
The second stage of adaptation for coastal communities is accommodation. The schemes employed in this stage allow for some degree of inundation (such as the building of flood-proof structures, and the use of floating agricultural systems).

Coastal retreat
The third and final stage of adaptation is retreat. Retreat can take various forms (like managed retreat, the building of temporary seawalls, or the monitoring of coastal threat to determine if and when evacuation is necessary).

The cost of inaction

While some adaptation strategies—for example, the construction of massive coastal defenses such as sea walls—could be expensive, the cost of inaction is arguably far greater in terms of lost lives and property. And most cost-accounting doesn't include collateral damages to coastal businesses, social institutions, ecosystems, and the environment (◀ p.124). For many island and low-lying regions, adaptations are urgently required; in the absence of adaptation, even the climate changes projected in mid-range scenarios could render these regions unlivable. Given business-as-usual emissions, there is no amount of adaptation that will maintain the resources—such as land and fresh water for farming—necessary for many of these island inhabitants to maintain their way of life and their unique cultures. Indeed, one Pacific island (Tuvalu) has already begun to plan for possible future evacuation to New Zealand. Just 1.0 m (3.3 ft) of sea level rise would displace more than 100 million people globally, underscoring the potential challenge of environmental refugeeism as massive populations are increasingly forced to flee inundated or otherwise uninhabitable coastal regions.

As with other climate change threats, ethical considerations arise from the disparity in wealth and resources between developed and developing countries (▶ p.206). Adaptation will naturally be more challenging for poorer nations, due to their less-developed adaptive capacity (▶ p.162) and their limited financial ability to fund costly engineering projects.

Keeping the water flowing

We know that the global demand for fresh water will rise as population grows. We also know that in many regions, increasing demand will coincide with a decreased water supply due to the impacts of climate change (◀ p.132).

The challenges ahead

Current water-management practices are unlikely to be adequate for addressing the new and additional challenges resulting from climate change. How will we alleviate both the stress of worsened water pollution on the environment and ecosystems, and the increased flood risk associated with more intense rainfall? How will we address the repercussions of diminished energy resources resulting from reduced river flow in many regions (◀ p.134)? And how will we tackle the problem of dwindling drinking and irrigation stores? Fortunately, there are changes in water-management practices that may help us with the daunting challenges ahead.

Adapting management practices

Communities can commit to making "no-regrets" refinements in water-management practices—that is, changes that will be helpful in dealing with the challenges posed by natural year-to-year variations in climate, regardless of whether or not human-caused global warming ultimately proves to be a threat. For example, in the western U.S., where there

Rolling sprinklers
More efficient irrigation methods may prove to be an effective means of adaptation in the face of diminished fresh water supplies.

is considerable year-to-year fluctuation in drought and flood conditions due to ENSO (◀ p.100), existing practices designed to deal with this variability could be exploited and refined to accommodate climate change impacts as well. Examples of adaptation procedures include the development of sea-water desalinization facilities, the expansion of reservoirs and rainwater storage facilities, and improvements in water-use efficiency and agricultural irrigation practices.

Planning for the future

Adaptation strategies are already being developed in regions such as North America, Europe, and the Caribbean in recognition of the potential for changes in precipitation patterns and water availability. In some cases, climate model predictions are being taken into account when adaptation procedures are being designed. However, predictions of regional changes in precipitation and drought patterns are still uncertain (◀ p.100). Consequently, so are the projected changes in river flow and water levels. As long as such uncertainties persist, it will remain difficult for water managers to develop optimal strategies. Nevertheless, being ready for more of what we have already seen in terms of year-to-year variability makes good sense, no matter what the future holds.

Ethical dimensions

The developing world has less access to the technologies required for effective adaptation to potential decreases in the fresh water supply. The greater vulnerability of the poor raises ethical dilemmas (▶ pp.206–207).

ADAPTATION OPTIONS IN THE FACE OF DECREASING FRESH WATER SUPPLY

Supply side	Demand side
Prospecting and extraction of water	Improvement of water-use efficiency by recycling groundwater
Increase of storage capacity by building reservoirs and dams	Reduction in water demand for irrigation by changing the cropping calendar, crop mix, irrigation method, and area planted
Desalination of sea water (via reverse osmosis systems)	Reduction in water demand for irrigation by importing products
Expansion of rainwater storage	Adoption of indigenous practices for sustainable water use (drawing upon local cultural knowledge in establishing efficient practices)
Removal of invasive, non-native vegetation from river margins	Expanded use of water markets to reallocate water to highly valued areas
Transport of water to regions where needed	Expanded use of economic incentives, including metering and pricing to encourage water conservation

A hard row to hoe

Climate change impacts on agriculture, livestock, and fisheries may jeopardize our ability to provide adequate food for a growing global population (◀ p.140). Are there adaptations we can make to protect ourselves from these impacts?

Getting ahead of the curve

Some agricultural options include changing crop varieties, locations, and planting schedules in response to changing seasonal temperature and precipitation patterns. These techniques may reduce some negative impacts and, in certain cases, even convert impacts from harmful to beneficial.

Adaptive practices could lead to increased crop yields in temperate latitudes, and potentially maintain current yields in tropical latitudes if warming is only moderate. If warming becomes high enough, however, the stress on water supplies may increasingly limit the benefits of adaptative strategies.

While adaptations can offset harmful impacts and even yield positive impacts, implementing them will require both a rethinking of governmental policies and the creation of new institutions to facilitate changes at the local level. It is therefore important that these measures be integrated into future economic development strategies.

Adaptation measures are not without some cost to communities and the environment. And implementation faces some obstacles. As crop yields begin to decrease in response to climate change impacts, there may be greater pressure on farmers to adopt unsustainable practices in an attempt to maximize short-term yields.

Winners and losers

There are ethical considerations that also come into play. Small farmers and subsistence farmers in tropical regions will be most vulnerable to climate change impacts due to their relative lack of access to institutions that can facilitate adaptation. Yet their contribution to greenhouse gas emissions is minimal.

Ironically, the farm industry in temperate regions such as the U.S., which is a major emissions contributor, may stand to benefit slightly from modest warming (◀ p.140), and is more likely to have access to any needed aid.

Wheat beats the heat
Wheat yields are projected to increase in extratropical regions like the U.S., but to decline in tropical regions. Appropriate adaptations could prevent the latter, as long as future warming levels are moderate.

CLIMATE CHANGE IMPACTS ON AGRICULTURE, LIVESTOCK, AND FISHERIES

WARMING RELATIVE TO CURRENT TEMPERATURES

+3.0–5.0°C (5.4–9.0°F)

+2.0–3.0°C (3.6–5.4°F)

+1.0–2.0°C (1.8–3.6°F)

Sub-sector	Region	Finding	Alleviation after adaptation
Prices and trade	Global	• Reversal of downward trend in wood prices • Agricultural prices: +10% to +40% • Cereal imports of developing countries: +10% to +40%	
Pastures and livestock	Low latitudes	• Strong production loss in swine and confined cattle	
Food crops	Low latitudes		• Maize and wheat yields reduced, regardless of adaptation; adaptation maintains rice yield at current levels
Pastures and livestock	Semi-arid	• Reduction in animal weight and pasture growth; increased frequenc of livestock heat stress and mortality	
Food crops	Global		• 550 ppm CO_2 (approx. equal to +2.0°C (+3.6°F) with no adaptation increases rice, wheat, and soy-bean yields by 17%
Prices	Global	• Agricultural prices: −10% to +20%	
Food crops	Mid to high latitude		• Adaptation increases all crop yields above current levels
Fisheries	Temperate	• Positive effect on trout in winter, negative in summer	
Pastures and livestock	Temperate	• Moderate production loss in swine and cattle	
Pastures and livestock	Semi-arid	• Increased frequency of livestock heat stress	
Food crops	Low latitudes		• Adaptation maintains yields of all crops above current levels; yields drop below current levels for all crops without adaptation
Food crops	Mid to high latitudes	• Crop growth less likely to be limited by length of growing seasons • No overall change in rice yield; regional variation is high	• Adaptation of maize and wheat increases yield by 10–15%
Pastures and livestock	Temperate	• Livestock grazing less likely to be limited by length of growing seasons; seasonal increased frequency of livestock heat stress	
Food crops	Low latitudes	• Without adaptation, wheat and maize yields reduced below current levels; rice yield is unchanged	• Adaptation of maize, wheat, and rice maintains yields at current levels
Pastures and livestock	Semi-arid	• No increase in productivity of plant growth; seasonal increased frequency of livestock heat stress	
Prices	Global	• Agricultural prices: −10% to −30%	

Part 5
Solving Climate Change

 Adaptation alone is unlikely to avert the most severe impacts of human-caused climate change. Instead, we must take action to mitigate the buildup of atmospheric greenhouse gases that are responsible for observed and projected global warming. Doing so will require that we reduce our reliance on fossil fuels by altering governmental policies and individual lifestyles. Important first steps we must take include forging cooperative relationships with other nations and rethinking how we, as a global community, can satisfy our energy requirements for key sectors of our economy—such as transportation, buildings, and agriculture—with minimal costs to planet Earth.

Key ideas

Technology & climate change

- There are two ways to mitigate against climate change: emitting less CO_2 and removing it from the atmosphere.

- The largest contributor to CO_2 emissions is energy supply followed by forestry, agriculture, industry, and transportation. In all of these sectors, there are opportunities to generate or use energy more efficiently.

- Other engineering approaches involve either removing greenhouse gases from the atmosphere or offsetting climate change by, for example, decreasing the amount of sunlight reaching Earth's surface.

Nations & individuals

- Our lifestyle choices—such as how we heat our homes and the forms of transportation we use—can also help to reduce greenhouse gas emissions.

- Tackling climate change will also require international cooperation. There are successful precedents for this, but several key nations have yet to make binding commitments to climate change agreements.

- The ethics of climate change— such as who should bear most responsibility and face the greatest risks—are often neglected.

- Although there are gaps in our understanding of climate change, this cannot justify inaction. A concerted human effort could bring many side benefits and has great potential to succeed.

Solving global warming

There are two ways to mitigate global warming: we can reduce or eliminate fossil-fuel carbon dioxide emissions, or we can remove the carbon dioxide from the atmosphere (▶ p.192). When we attempt the latter strategy, and remove CO_2, it is referred to as carbon capture and storage (CCS) or carbon sequestration.

Thankfully, there are no insurmountable technological or scientific reasons why we can't employ either strategy: emission reduction or carbon capture and storage. The only barrier is society itself. Although many countries are attempting to reduce them, emissions continue to grow, and atmospheric CO_2 levels are climbing at rates that exceed previous projections. This is largely because there are few economic incentives for emission reduction, especially in the two largest emitting nations, the United States and China.

Carbon costs

A potential solution to global warming is to translate the social cost of carbon (◀ p.156) into a carbon cost that is paid by the consumer who emits. A carbon cost is an amount that consumers must pay (as a tax or as part of an emission permit

Carbon reduction

These graphs show by how much emissions of CO_2 would be reduced by each sector if they were taxed at the specified rate per metric ton. For example, if emission of 1 metric ton of carbon is taxed at U.S. $20, the forestry sector will potentially reduce emissions by just over 1 Gt CO_2 eq. If the carbon cost or tax is raised to $100, the forestry sector may be incentivized to reduce emissions by more than 4 Gt CO_2 eq. (The **CO_2 equivalent** is a unit that expresses the combined impact of multiple greenhouse gases in terms of the impact of an equivalent amount of CO_2.)

REDUCTION POTENTIAL AT 3 DIFFERENT CARBON COSTS (U.S. $ per metric ton)

Energy supply — GT CO_2 EQUIVALENT PER YEAR: $20, $50, $100

Transport: $20, $50, $100

Buildings: $20, $50, $100

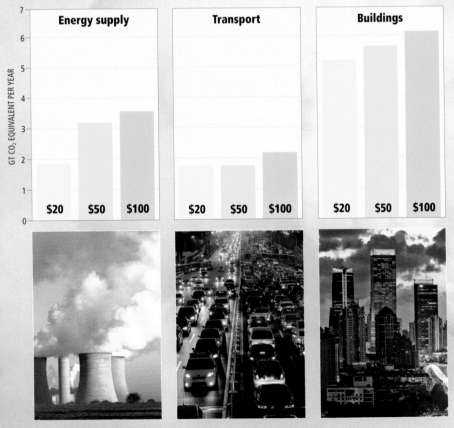

exchange) for the emission of one metric ton (1.1 U.S. tons) of CO_2. Taxation would not only reduce consumption, but also provide an incentive for the development of non-carbon energy sources.

Emission reduction potential

Mitigation efforts by necessity must span many sectors of the economy, from energy supply, transport, and buildings, to industry, agriculture, forestry, and waste management. The largest potential for emission reductions can be found in some unexpected places, depending on whether we credit reductions at the point of emission or at end-use. If point of emission is credited, then the largest reductions would be in the energy supply sector.

However, if we look at end-use, then the buildings sector rises in importance (see graphs below). In all sectors, emissions are projected to decrease as the carbon cost increases. Note, though, that with the exception of the forestry sector, investing larger and larger amounts of money in carbon emission reductions leads to smaller incremental gains—the so-called law of diminishing returns.

In the pages that follow, we will investigate how emission reductions might be achieved in each of the major economic sectors, and consider, in turn, alternative **geoengineering** fixes such as carbon sequestration and the reduction of incoming sunlight.

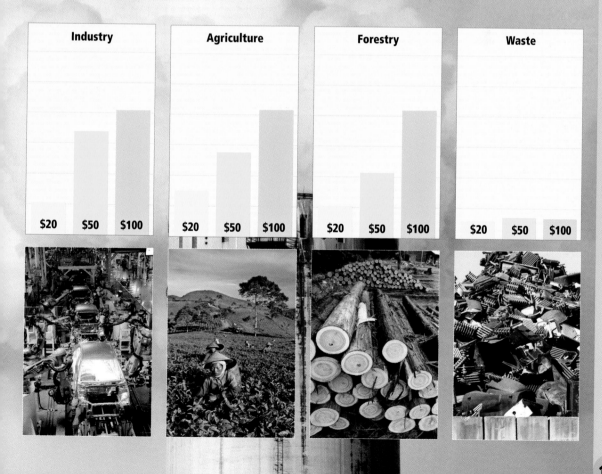

Industry — $20 $50 $100

Agriculture — $20 $50 $100

Forestry — $20 $50 $100

Waste — $20 $50 $100

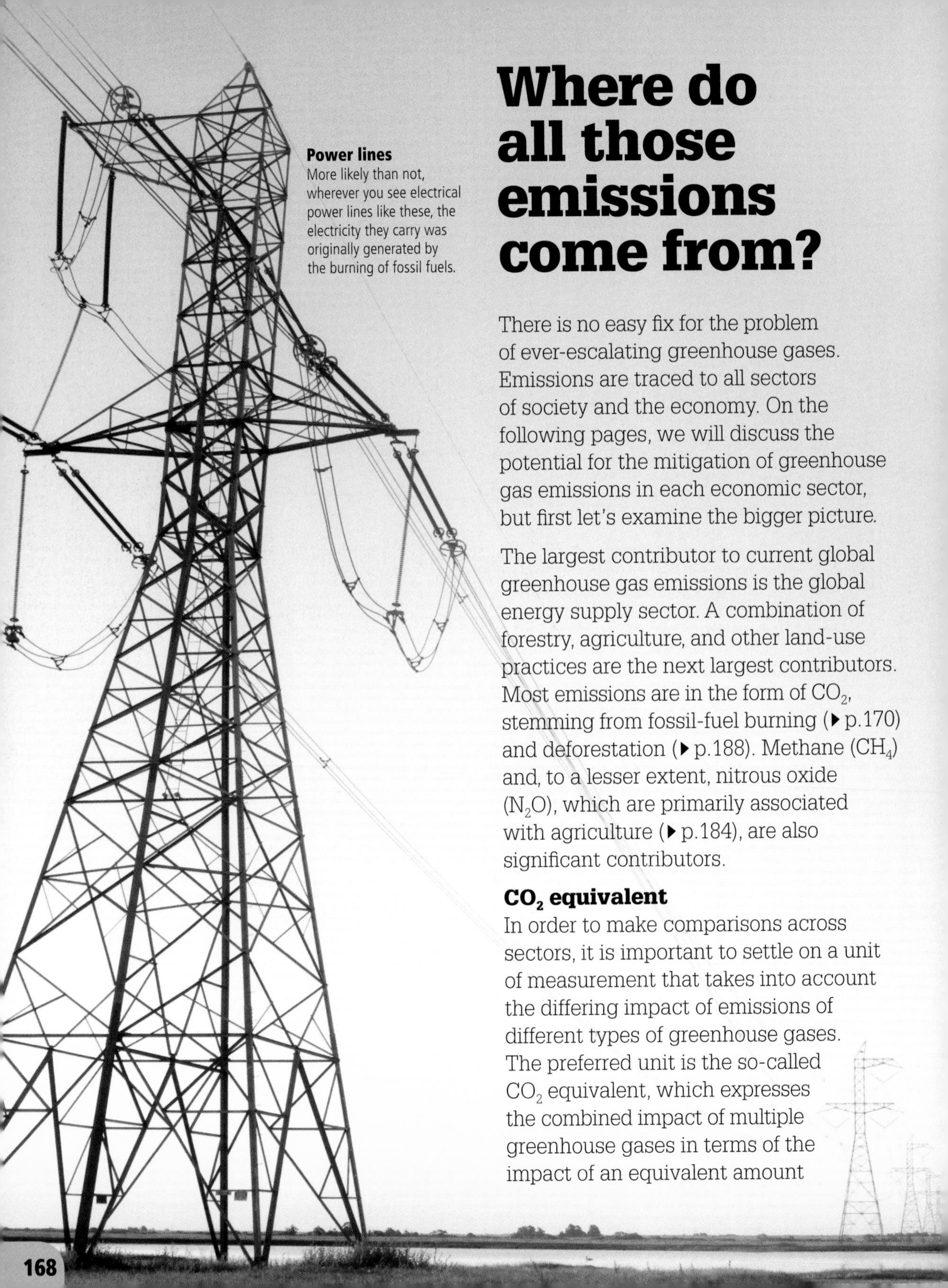

Power lines
More likely than not,
wherever you see electrical
power lines like these, the
electricity they carry was
originally generated by
the burning of fossil fuels.

Where do all those emissions come from?

There is no easy fix for the problem of ever-escalating greenhouse gases. Emissions are traced to all sectors of society and the economy. On the following pages, we will discuss the potential for the mitigation of greenhouse gas emissions in each economic sector, but first let's examine the bigger picture.

The largest contributor to current global greenhouse gas emissions is the global energy supply sector. A combination of forestry, agriculture, and other land-use practices are the next largest contributors. Most emissions are in the form of CO_2, stemming from fossil-fuel burning (▶ p.170) and deforestation (▶ p.188). Methane (CH_4) and, to a lesser extent, nitrous oxide (N_2O), which are primarily associated with agriculture (▶ p.184), are also significant contributors.

CO_2 equivalent

In order to make comparisons across sectors, it is important to settle on a unit of measurement that takes into account the differing impact of emissions of different types of greenhouse gases. The preferred unit is the so-called CO_2 equivalent, which expresses the combined impact of multiple greenhouse gases in terms of the impact of an equivalent amount of CO_2. The CO_2 equivalent is typically measured in either megatons (millions of metric tons) or **gigatons** (billions of metric tons) of CO_2 (abbreviated as Mt/Gt CO_2 eq).

Who emits?

Although the energy supply sector is currently responsible for the largest emissions (nearly 17 Gt CO_2 eq annually), other sectors such as agriculture and forestry (around 12 Gt), industry (around 10 Gt), and transportation (around 7 Gt) continue to contribute substantially to global greenhouse gas emissions.

Over the past four decades, CO_2 emissions from fossil-fuel burning alone have roughly doubled from 16 Gt CO_2 eq to about 32 Gt CO_2 eq annually. Fossil-fuel burning now accounts for approximately two-thirds of total greenhouse gas emissions.

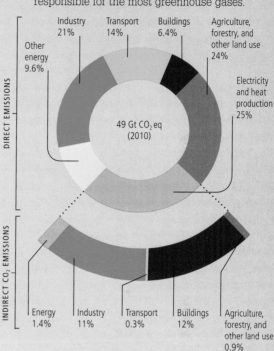

GREENHOUSE GAS EMISSIONS BY SECTOR IN 2010

Electricity generation and agriculture/forestry are responsible for the most greenhouse gases.

DIRECT EMISSIONS

Other energy 9.6%
Industry 21%
Transport 14%
Buildings 6.4%
Agriculture, forestry, and other land use 24%
Electricity and heat production 25%

49 Gt CO_2 eq (2010)

INDIRECT CO_2 EMISSIONS

Energy 1.4%
Industry 11%
Transport 0.3%
Buildings 12%
Agriculture, forestry, and other land use 0.9%

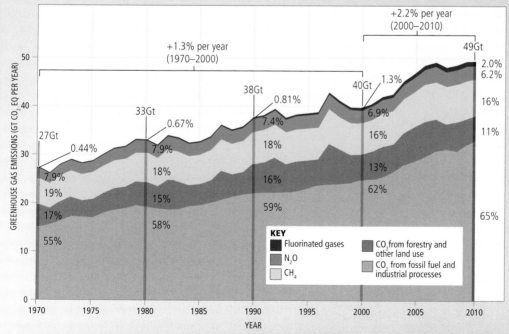

+1.3% per year (1970–2000)

+2.2% per year (2000–2010)

49Gt
2.0%
6.2%
40Gt
1.3%
6.9%
16%
38Gt
0.81%
7.4%
18%
16%
62%
33Gt
0.67%
7.9%
18%
15%
59%
27Gt
0.44%
7.9%
19%
17%
55%
58%
16%
11%
65%

GREENHOUSE GAS EMISSIONS (GT CO_2 EQ PER YEAR)

50
40
30
20
10
0

1970   1975   1980   1985   1990   1995   2000   2005   2010
YEAR

KEY
- Fluorinated gases
- N_2O
- CH_4
- CO_2 from forestry and other land use
- CO_2 from fossil fuel and industrial processes

TRENDS IN GREENHOUSE GAS EMISSIONS OVER TIME

Carbon dioxide (CO_2) from fossil-fuel burning and deforestation continue to make up the bulk of greenhouse gas emissions.

Keeping the power turned on

The consumption of fossil fuels, mainly coal and natural gas, generates much of the world's energy supply—the energy we use for electricity generation and heating (oil is primarily used for transport). The energy sector is the single largest source of greenhouse gas emissions, responsible for over 25% of all worldwide emissions. The primary culprit is CO_2, though methane released during fossil-fuel processing is also significant. Despite recent international efforts to develop and use non-carbon and renewable energy sources, and the introduction of policies, such as carbon trading and higher energy prices, emissions have increased substantially in recent years. From just 1990 to 2010, annual energy-related emissions nearly doubled from 9 to 17 Gt CO_2 eq (◄ p.168).

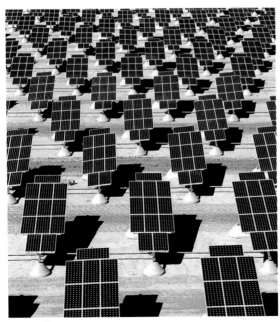

Sunseekers
These 70,000 solar panels are part of a solar photovoltaic array that generates 15 megawatts of solar power for Nellis Air Force Base in Nevada.

How can we stabilize emissions?

Without widespread governmental action, energy-related emission rates will likely rise an additional 50% in the coming decades. As emissions continue and their rates increase, stabilizing greenhouse gas concentrations will become ever more challenging. One common misconception is that the "Peak Oil" phenomenon (the projected impending depletion of readily available petroleum reserves) will solve the fossil-fuel emissions dilemma. However, even if oil wells run dry, the primary sources for the energy sector—coal and natural gas reserves—could last for centuries. In reality, meeting the rising global demand for energy supply while simultaneously slowing the rate of fossil-fuel emissions will require a combination of tools. We need to strive for greater efficiency in power generation, an increased use of carbon-free (e.g., nuclear, solar, and wind) or carbon-neutral (e.g., **biofuels**) energy sources, and the continued development of carbon capture and storage (CCS) technologies.

Energy alternatives

Carbon-free and carbon-neutral energy sources each have their merits and weaknesses. Increased use of nuclear energy (which currently accounts for about 7% of the global energy supply) is limited by a number of factors, including the restricted availability of uranium, security considerations, safety issues, and limited public support. While renewable energy sources such as solar, wind power, and

WORLD PRIMARY ENERGY CONSUMPTION BY FUEL TYPE

Despite modest increases in the use of renewable energy resources in recent decades, fossil-fuel sources (gas, coal, and oil) continue to supply the greatest share of the world's energy.

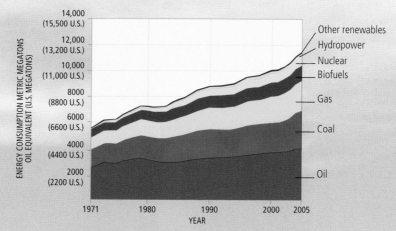

geothermal are currently minor contributors to global energy supply, government incentives could encourage development and increased efficiency. However, the localized and variable availability of these sources are obstacles to widespread use in major urban centers. The use of biofuels—such as wood, sugarcane, vegetable oil, and even dung for heating and cooking—which currently accounts for more than 10% of global energy consumption, could be modestly expanded (▶p.186). The increased use of hydropower, another renewable energy source, faces opposition due to potential environmental threats posed by major damming projects. Moreover, most rivers that can be dammed already have been (and the remaining rivers are largely protected). In many regions, the viability of **hydropower** may also be threatened by climate change itself, i.e., shifting precipitation patterns (◀p.132).

Wind power
Wind turbine generators turn at Scroby Sands off the coast of Norfolk, U.K. Increased efficiency of wind farms could help stabilize energy-related emission rates.

(Cont.)

No easy answers

(Cont.)

There is clearly no easy way to meet the world's rising energy demands in a climate-friendly manner. All options need to be considered, at least in the short term. While the developed world has the highest per-capita energy demand, the most rapid growth in energy use is taking place in developing countries such as India and China. Decreasing fossil-fuel emissions will require cooperation between the developed and developing world (▶ p.200).

Can we frack our way out?

An important recent development is a dramatic trend away from coal toward (currently cheaper) natural gas for power generation. Whether this trend is favorable from a carbon emissions standpoint isn't entirely clear yet. While natural gas (largely methane) nominally produces only about 50% as much CO_2 as coal per watt of power generated, there are still open questions regarding the magnitude and extent of so-called fugitive methane that may be released in the process of hydraulic fracturing (fracking). Since

Fracking
Rigs drilling for shale gas are becoming common across the U.S. Shale gas now provides around 20% of the country's natural gas supply.

methane is an even more potent greenhouse gas than CO_2 (◀ pp.26–27), these emissions could at least partly offset any greenhouse gas reductions that might be achieved through substitution of natural gas for coal.

Capturing Earth's heat
Iceland sits on the Mid-Atlantic ridge, a region of volcanic upwelling that is slowly widening the Atlantic Ocean. Geothermal power plants tap into the volcanic heat to produce electricity in a manner that is sustainable, reliable, non-polluting, and environmentally friendly.

METRIC TONS	0	1.5	3.0	4.5	6.0	>
U.S. TONS	0	1.6	3.3	5.0	6.6	>

TONS OF OIL EQUIVALENT (TOE) PER CAPITA

WORLD ENERGY CONSUMPTION BY REGION

While energy consumption is increasing in regions such as China and India, per-capita energy consumption continues to be highest in the developed world.

On the road again

We rely almost exclusively on petroleum-based fuels, such as gasoline, for transport. This fuel use results in emissions of about 7 Gt CO_2 eq per year (◀ p.168), and it is responsible for 14% of worldwide greenhouse gas emissions. Road vehicles produce about 75% of this total. Emissions in the transport sector are increasing at an even faster rate than those in the energy sector. The greatest growth occurred in the area of freight transport (primarily by trucks for overland freight, and ships and airplanes for international transport). The rate of transport emissions is projected to increase even further over future decades, fueled by continued global economic growth and population increase.

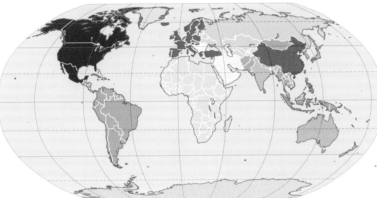

PROJECTION OF TRANSPORT ENERGY CONSUMPTION BY REGION 2000–2050

Developing nations such as China and India are dramatically increasing their share of transport-related greenhouse gas emissions.

-12 -10 -8.0 -6.0 -4.0 -2.0 0 2.0 4.0 6.0 8.0

DIFFERENCE IN THE PERCENTAGE SHARE OF OVERALL TRANSPORT ENERGY CONSUMPTION, BY REGION, BETWEEN 2000 AND 2050 (PROJECTED)

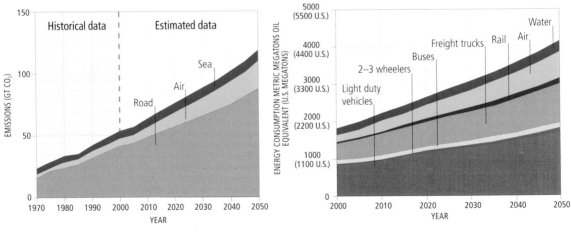

HISTORICAL AND PROJECTED TRANSPORT EMISSIONS BY MODE 1970–2050

Land-based transport modes will continue to dominate greenhouse gas emissions in the decades ahead, but air travel is projected to make increasing contributions.

PROJECTION OF TRANSPORT ENERGY CONSUMPTION BY MODE 2000–2050

Much of the future increase in transport-related energy consumption is likely to come from a combination of personal ("light duty") vehicles and freight trucks.

New fuels
This fuel pump in Spain serves up fuel consisting of 30% biodiesel and 70% conventional petrodiesel. Biofuels are most commonly used as mixtures with conventional fuels like petrodiesel (as here) and gasoline, but they can also be used in pure form to power vehicles.

While most people in developing countries do not own personal vehicles or have any access to motorized transportation, this situation is changing. This is particularly true in developing nations like China, where personal automobile ownership has skyrocketed in recent years. In Beijing, car ownership over the past 10 years has increased from only 2 to about 40 per 100 households. In the absence of a radical shift from current practices, transport-related carbon emissions may nearly double in the next few decades.

The potential of the "Peak Oil" phenomenon (◀ p.170) to slow the future rate of growth has been overstated. Even if conventional oil fields were depleted in coming decades, or if drilling were prohibitively expensive, sources such as oil shales and tar sands could provide many additional decades of petroleum reserves. As coal liquification technology advances, coal could potentially satisfy the transport sector's rising fuel demands.

Emerging **fuel-cell technology** has often been cited as a potential solution. However, the use of fuel cells alone for passenger vehicles could simply shift the energy burden from the transport sector to the energy sector, with no net decrease in emissions. This is because fuel cells, which convert fuel energy into electricity in a manner similar to a battery, have to be recharged, and the energy to do that has to come from somewhere.

The quest for fuel efficiency
Meeting rising global transport-sector energy demands while slowing the rate of fossil-fuel emissions will require a combination of greater fuel efficiency and increased use of carbon-free or carbon-neutral technologies. In the short term, this can be accomplished with more fuel-efficient vehicles, such as gasoline/electric hybrid cars and clean diesel vehicles, and through increased use of certain types of biofuels as gasoline additives or substitutes. In one to two decades from now, biofuels could satisfy 5–10% of the total transport energy demand. Technological innovations and improved air traffic management could result in better fuel efficiency in the aviation sector. Increased reliance on trains, buses, and other public transport, car-pooling, and non-motorized transportation (such as cycling and walking) could also help to curb emissions.

(Cont.)

GROWTH OF ELECTRIC VEHICLES

Since about the beginning of this decade, there has been a dramatic increase in the number of electric vehicles sold in the U.S., from almost none at the start of 2011 to more than a quarter of a million by Fall 2014. Approximately half were all-electric vehicles, and half were plug-in hybrids.

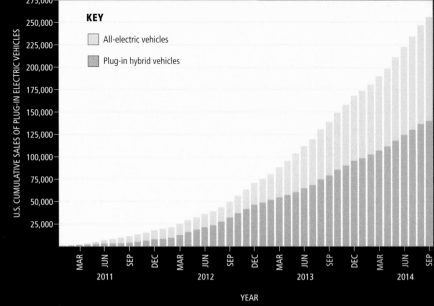

KEY

☐ All-electric vehicles

☐ Plug-in hybrid vehicles

U.S. CUMULATIVE SALES OF PLUG-IN ELECTRIC VEHICLES

YEAR

"No regrets"

In many cases, measures to reduce fossil-fuel consumption constitute "no regrets" strategies. For example, using our automobiles less has the added benefit of reducing traffic congestion and improving air quality. And decreased gasoline consumption results in the national security bonus of reducing our reliance on volatile regimes. Improved vehicle efficiency measures also lead to savings in fuel expenditures—savings that can be invested in other areas of the economy.

Long-term plans

Emerging technologies may allow for further reductions in transport-related greenhouse gas emissions in coming decades. New types of biofuels, electric and hybrid vehicles with more powerful and longer-running batteries, and more efficient aircraft can all contribute to long-term emission-reduction goals. In European countries such as Norway, market shares of electric and plug-in hybrid vehicles have reached double-digit

percentages. By contrast, in the U.S., market share is less than 1% but growing rapidly, led by successful brands like the Nissan Leaf and the high-end, sporty Tesla Model S. To yield substantial carbon emission reductions, the power grid used to charge electric vehicle batteries must become less reliant on fossil-fuel energy sources.

Who can help?

Public policy measures can aid the mitigation of transport-related emissions in a number of ways. This is particularly true for nations or communities that are still in the process of establishing transportation systems. For example, careful urban planning and land-use regulations can reduce commuting distances, provide better access to public transport, and make public transport a more appealing option. Governments can establish, enforce, and, where necessary, raise mandatory fuel-economy standards. Appropriate taxation and fees can encourage the use of efficient vehicles. Some obstacles to the success of

Refueling station for electric vehicles
In an environmentally friendly future, we could see stations like these in Paris, France, replacing gas pumps as electric vehicles replace today's inefficient gas-guzzlers.

these measures are the persistent preference of many consumers for vehicles with poor fuel efficiency, targeted corporate advertising campaigns that reinforce these preferences, and lobbying efforts by car companies to keep fuel efficiency standards low. Any substantive reduction in the transport sector's fuel consumption will require not only proactive government policies, but also greater corporate accountability and personal acceptance of responsibility by individual citizens.

Building green

The commercial and residential buildings sector is a large emitter of greenhouse gases. Including its energy use, the buildings sector becomes one of the largest emitters of CO_2, emitting approximately 19% of total CO_2 equivalent per year in 2010. Including the energy-consumption-related emissions is important as we consider the climate impact of approaches taken to promote energy efficiency in the buildings sector. Happily, many of the approaches taken to reduce emissions in this sector are technologically mature and provide benefits in addition to reducing emissions.

There are two basic ways the buildings sector can reduce its **carbon footprint**:

- Reduce energy consumption in construction and building operation
- Switch to low-carbon or carbon-free energy sources

Here we focus on the first of these strategies; alternative energy sources are addressed in "Keeping the power turned on" (◄ p.170).

Reducing energy consumption

The green building movement encourages efficiency in the design, construction, operation, and demolition of buildings, with the goals of enhanced human health and reduced impact on the environment. In the U.S., this movement is known as the Leadership in Energy and Environmental Design (LEED). Certification points are awarded by LEED for **sustainability** and efficiency, as well as for the optimization of energy performance and the use and reuse of recyclable materials. A successful LEED building takes into account such

Taipei 101
Located in Taipei, Taiwan, Taipei 101 is 509 m (1670 ft) tall and is one of the world's most technologically advanced and "green" supertall skyscrapers, with an **LEED platinum rating** (the highest rating possible).

important concerns as the availability of public transportation to the building, habitat preservation, and indoor environmental quality.

Reducing energy use in new buildings means reducing the heating and cooling loads. This can be accomplished through passive solar design (taking best advantage of available solar energy) and better insulation. High-efficiency lighting, appliances, and heating and cooling systems; high-reflectivity building materials; and multiple glazing in windows can also markedly reduce a building's emissions. These systems can be optimized using building automation systems that monitor all essential functions and occupancy of the building. Zero net energy building is now achievable, using a combination of these best practices and active solar or wind energy.

Green renovation

Energy savings in new construction can exceed 90%, which bodes well for the future. However, buildings have long lifetimes, so most buildings in existence today will still be in use in 2030. This means that close attention must be paid to the renovation of existing buildings, because that is where most building-related emissions reductions will be made. Two- to four-fold reductions in energy use can be achieved through renovation of existing buildings.

Although these so-called "green" renovation techniques can involve considerable costs up-front, there are economic savings in the long term associated with energy-use reduction, and a number of co-benefits as well, including improved indoor air quality. Green building can also create jobs and new business opportunities, which in turn enhances economic competitiveness and energy security.

To enhance the lure of these long-term benefits, some measure of government intervention may be necessary. Appliance efficiency standards, new building codes, mandatory labeling and certification, energy-efficiency quotas, and tax benefits for green construction are all options. Most of these mechanisms have a high cost-effectiveness, and in many cases benefits can be realized without costs.

Stemming the rising greenhouse gas emissions from the buildings sector will require a strong political commitment to green construction. This may take the form of governmental monitoring and the enforcement of codes and regulations. In the end, though, if the green building movement is successful, the interior space where we spend much of our time will be healthier and more comfortable, and have a minimal impact on the global environment.

David L. Lawrence Convention Center
Situated in downtown Pittsburgh, Pennsylvania, this convention center was the first LEED-certified convention center in North America, with both platinum and gold certifications (the two highest ratings).

Reducing CO$_2$ pollution

The image of a factory belching out smoke is ingrained in our minds as the height of environmental pollution. Although many industries have taken significant steps to reduce pollution, due to its intense use of energy, the industry sector is still a major source of CO$_2$ and other greenhouse gases. And that source is growing: from 6 Gt CO$_2$ eq in 1971 to nearly 10 Gt CO$_2$ eq in 2010 (◀ p.168). In stringent scenarios where warming is capped at 3°C (5 °F) above pre-industrial levels, there will be increasingly severe reductions needed by 2050 and 2100 to stabilize the warming. The carbon cost of many of these industry-specific improvements is relatively

CARBON CAPTURE AND STORAGE (CCS)

Carbon dioxide can be captured before it is released into the atmosphere and transferred underground via pipelines. Possible repositories include coal and salt beds, depleted oil and gas reservoirs, and saline aquifers.

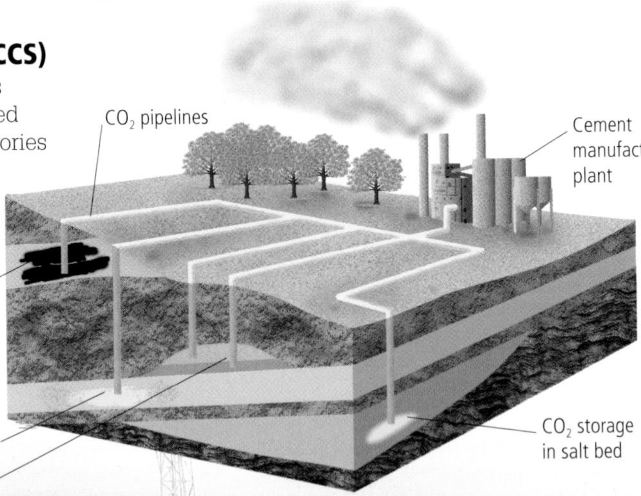

CO$_2$ pipelines

Cement manufacturing plant

CO$_2$ storage in coal beds

CO$_2$ storage in saline aquifer

CO$_2$ storage in depleted oil and gas reservoirs

CO$_2$ storage in salt bed

Back to where it came from

Each year, the Norwegian oil company Statoil is injecting about 1 Mt of CO$_2$ 1 km (0.6 miles) below the seafloor at its Sleipner West field in the North Sea.

INDUSTRY	CURRENT INTENSITY OR EMISSIONS	TECHNOLOGY AND INTENSITY	TARGET INTENSITY
Cement Tons of CO_2 per ton of cement	0.8	Current: 0.5–0.6 CCS: 0.1–0.2	0.35–0.6 (2030) 0.25–0.4 (2050)
Steel Tons of CO_2 per ton of steel	2.2	Current: 0.2–1.6 CCS: 0.6–0.8	0.9–1.4 (2030) 0.4–0.8 (2050)
Chemicals Gt CO_2 eq direct emissions	1.8	Current: 1.3–1.7 CCS: 0.4–1.7	1.6 (2030) 1.3 (2050)
Paper Tons of CO_2 per ton of paper	0.58	Current: 0.48–0.52 CCS: 0.05–0.14	0.28 (2030) 0.18 (2050)

affordable (◀ p.166), making this sector an attractive one for targeted reductions.

Industrial CO_2 emission is increasingly becoming an issue for developing nations. In 1971 only 18% of industry-related CO_2 emissions came from developing countries, but by 2004 their share had risen to 53%. This shift reflects both the growth of industry in developing countries and the movement toward improved energy efficiency in developed countries.

INDUSTRY TARGETS

Each of the industries in this table is a carbon emitter. To achieve 450 parts per million (ppm) CO_2 by 2100, industry must reduce its CO_2 emissions and carbon intensity to the levels shown in the far right column. One strategy to mitigate CO_2 emissions is carbon capture and storage (CCS) to prevent carbon being released into the atmosphere. Currently available and CCS technology could achieve the intensities of the middle column.

Industrial mitigation

There are numerous opportunities for mitigation of greenhouse gas emissions in industry, in part because many factories use old and inefficient processes. Retrofitting these factories—replacing electric motors and boilers, using recycled materials for fuel, and fixing leaks in furnaces and air and steam lines—could go a long way toward limiting future carbon emissions. Carbon capture and storage is another promising strategy to help realize industrial reductions—which will be needed to achieve the goals for 2100.

The water-energy nexus

Transportation and electricity generation consume a considerable amount of water. While some water is used in the extraction and processing of fossil fuels for transportation, biofuel production (◀ p.170) is extremely water intensive because of irrigation needs. And although wind and solar photovoltaic energy generation require little water, other alternative electric energy sources such as geothermal and hydroelectric are highly water intensive. So, the substitution of alternative energies for fossil fuels can certainly reduce carbon consumption but

NOT ALWAYS GREENER

Producing and consuming energy places a demand on water supplies, and renewable biofuels, such as soy and corn, consumes considerably more water than the equivalent energy production from fossil fuels.

could also strain the world's water supplies. The water-energy nexus is in reality a water-energy-food-forest-climate nexus. Decisions about energy production—for example the extent to which renewable biofuels should substitute for fossil fuels—impacts not only water supplies but also the ability of our agricultural system to meet fundamental food needs. In 2014, biofuels consumed 40% of the U.S. corn crop; 45% went to feed livestock, and only 15% was used directly to feed people. Overall, agriculture utilizes 80% of the U.S. water supply, but electricity generation is also significant, consuming about 5% of the U.S. water supply.

As nations look to alternatives to fossil fuels for energy production, they must consider the extra demands a switch to renewables might place on water. Water scarcity is a growing concern in many regions, so a focus on water-efficient alternative energy production seems prudent.

GALLONS OF WATER USED TO PRODUCE 1 MILLION BTU'S

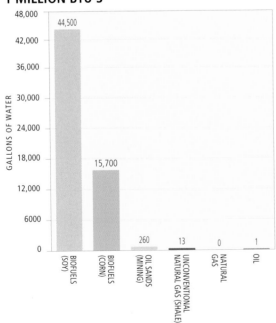

WATER CONSUMPTION OF AN 18,000 BTU AIR CONDITIONER OVER A WEEK

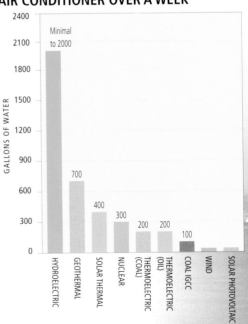

Water concerns
Electricity cables span a section of Shihwaho Lake in South Korea. Economic growth increases the demand for electricity, and countries need to be aware of the demand this can in turn place on water supplies.

Greener acres

Nearly half of Earth's land surface is used for farming (crops and grazing), so it should come as no surprise that the agriculture sector is a significant contributor to global greenhouse gas emissions. Farming and agriculture are responsible for annual emissions of about 6 Gt CO_2 eq (◀ p.168), or about 12% of worldwide greenhouse gas emissions, roughly equal to the contribution from transport. Agricultural emissions have stabilized in recent years, and an increasingly large share comes from the developing world, which is now responsible for about 75% of worldwide agricultural emissions. Interestingly, net CO_2 emissions from agriculture are negligible; plants produce CO_2, but they consume it at about the same rate. The main agricultural emission is methane (CH_4), produced by microbes that thrive in environments such as rice paddies and the stomachs of ruminants like cattle, oxen, and sheep. Nitrous oxide (N_2O) from manure and other fertilizers is another agriculture-generated greenhouse gas.

Agricultural mitigation

Carbon is absorbed by and stored in living plants on Earth. Farming and grazing lands thus represent large potential carbon stores (places where carbon is sequestered away from the atmosphere). The land's ability to store carbon has decreased significantly over time, due to heavy-handed agricultural practices such as overtilling soils. Consequently, the greatest potential for agricultural mitigation lies not in reduction of emissions themselves, but improved management of agricultural lands, which will restore their ability to sequester CO_2. Positive management practices include reducing soil tillage and restoring carbon-absorbing organic soils. More Earth-friendly practices that restore degraded land, such as converting aging crop land to grassland, can similarly aid in mitigation.

Of course, it is also a good idea to decrease agricultural emissions. Improved management practices can play a key role in this endeavor. For example, more efficient fertilizer delivery methods can minimize nitrous oxide emissions. Rice paddies can be better managed to reduce methane production. And using alternative feeds will result in less methane production by ruminants. For example, recent experiments show that adding certain types of food-industry byproducts, such as cooking fat, to cattle feed reduced their methane production.

Harvesting rice
If farmers, like these rice farmers in Guilin, China, take good care of their farm lands, they can dramatically increase the land's ability to sequester CO_2 from the atmosphere.

GLOBAL MITIGATION POTENTIAL OF VARIOUS AGRICULTURAL MANAGEMENT PRACTICES BY 2030

Better management of crop lands, grazing lands, and soils can decrease net greenhouse emissions by allowing land to more effectively sequester atmospheric carbon dioxide. Better farming practices can also reduce methane production, another important contributor to greenhouse gas emissions. The negative values on the graph indicate that rather than mitigating, the action in question is adding to current emissions.

KEY
- Nitrous oxide (N_2O)
- Methane (CH_4)
- Carbon dioxide (CO_2)

MEGATONS CO_2 EQUIVALENT PER YEAR

Categories:
- CROP LAND MANAGEMENT
- GRAZING LAND MANAGEMENT
- RESTORE CULTIVATED ORGANIC SOILS
- RESTORE DEGRADED LANDS
- RICE MANAGEMENT
- LIVESTOCK
- BIOFUEL
- WATER MANAGEMENT
- SET-ASIDE, LAND USE CHANGE & AGROFORESTRY
- MANURE MANAGEMENT

» (Cont.)

Changing our diet

(Cont.)

Because livestock are resource intensive, switching from diets heavy in meat consumption to more vegetarian diets may substantially decrease worldwide carbon emissions (▶p.198). A co-benefit of economic development efforts could involve encouraging such dietary shifts where appropriate.

Biofuel options

There is also potential for mitigation in agricultural production of biofuels. Burning biofuels does not lead to any net increase in greenhouse gas concentrations because the carbon that is released was just recently removed from the atmosphere, and has only been stored in plants for a few months. For every gram of CO_2 released when a biofuel is burned, a gram was removed from the atmosphere by photosynthesis just a short time ago. This balance is why biofuels are considered "carbon neutral." Crops and agricultural residues can be used either as crude biofuel energy sources (for example, burned for heat), or they can be chemically altered to yield more efficient biofuels such as ethanol or biodiesel.

Corn, one of the most widely grown cereal crops, can be readily converted into ethanol. However, there are at least two problems that limit prospects for the widespread use of corn-based ethanol as a fuel. First, there are troubling ethical considerations associated with the prospect of trading food for energy in this way, when starvation and malnutrition still afflict large numbers of people, especially in developing countries.

ESTIMATED MITIGATION POTENTIAL IN THE AGRICULTURAL SECTOR BY 2030

Note that South America, Southeast Asia, and Indonesia have the greatest mitigation potential.

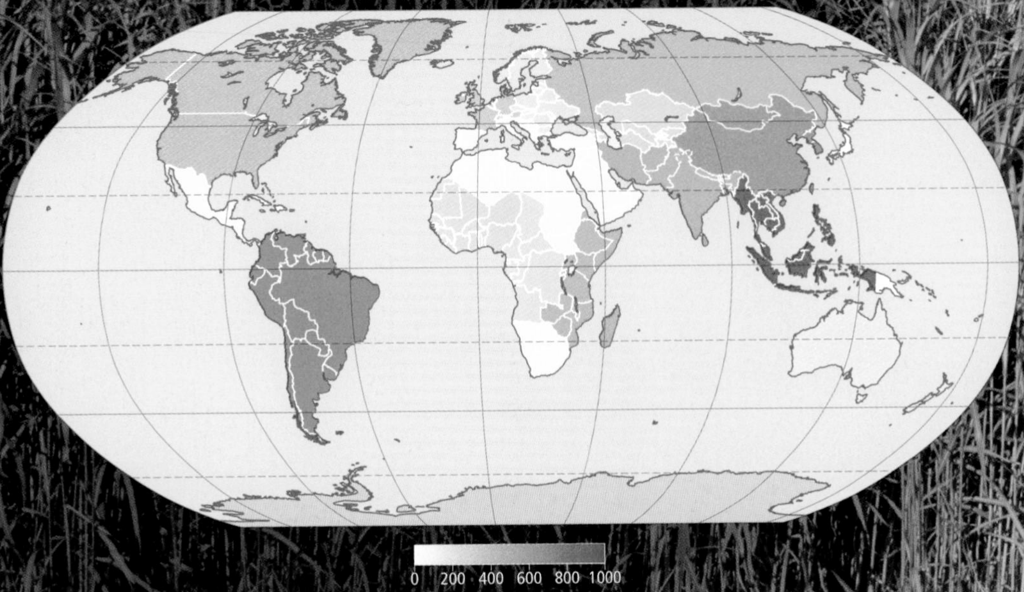

0 200 400 600 800 1000

ESTIMATED MITIGATION POTENTIAL (MT CO₂ EQ/YR)

Secondly, and perhaps more practically, the processes used to convert corn to ethanol are not very efficient; they require a considerable input of energy and water, limiting net gains. Researchers are currently seeking alternative pathways for ethanol production.

Cellulosic ethanol, which can be derived from agricultural products such as switchgrass, could yield larger net gains in the future. Switchgrass is a tall grass that grows naturally on the prairies of North America. Research has demonstrated that it has the potential to yield biofuel more efficiently than corn. In comparison with corn, switchgrass is more hardy with respect to soil and climate conditions, is perennial, and requires far less fertilizer and herbicide. And, it is not a food crop!

Uncertainties

Most (approximately 70%) of the potential for mitigation in the agricultural sector lies in the developing world, specifically in China, India, and much of South America. It is challenging to determine precisely what the mitigation potentials of various options are. Key uncertainties include the willingness of governments to promote and support mitigation practices, and the willingness of individual farmers to adopt preferable management practices. Also uncertain is the continued effectiveness of various mitigation strategies in the face of escalating climate change, growing populations, and evolving technology.

Switchgrass field
As a biomass crop for producing ethanol, switchgrass may yield greater mitigation benefits than corn.

Forests
Source or sink for atmospheric CO_2?

Long before humans were burning fossil fuels, we were contributing to the buildup of carbon dioxide in the atmosphere. The practice of deforestation has accompanied human settlement and agriculture across the globe. The combustion of timber for energy and the gradual decay of lumber used in construction both release CO_2 into the atmosphere. In 2010, the forestry sector (including land use other than agriculture) emitted roughly 11% of the total greenhouse gases released to the atmosphere.

Carbon uptake
During the pre-industrial era, forest-clearing and wood burning were common practices in Europe and the U.S. Recently, however, previously cleared agricultural lands have returned to forests in these regions. As a result, forest-related CO_2 emissions have declined over the last several decades (see graph opposite) and reforested lands have now become carbon sinks.

Now the developing world is repeating history, aggressively cutting down and burning trees. Deforestation in tropical South and Southeast Asia, Africa, and South America has recently accelerated. Between 2000 and 2012, an area roughly the size of Alaska and Texas combined was lost to deforestation.

This new carbon influx from tropical deforestation has been partially offset by the reforestation uptake in Europe, America, and elsewhere. Current emissions from deforestation amount to nearly 4.5–5.5 Gt CO_2 eq (◀ p.168) per year. Encouragingly, these emissions have declined in recent years. However, the decline is the result of decreased deforestation in high-income countries; low-income countries have increased deforestation in recent years. On the flip side, it is estimated that approximately 3.3 Gt CO_2 eq per year are taken up through reforestation.

Obviously, the most expeditious way to reduce CO_2 emissions from the forestry sector is to prevent deforestation. But the developed world must keep in mind its own history.

Reforestation potential
What about mitigating via reforestation efforts? The efficacy of reforestation is contingent on favorable climate conditions. Deforested soils tend to dry out, and they are often low in nutrients because most of the ecosystem nutrients were stored in the harvested trees. For this and other reasons, reforestation of tropical rainforest has often been unsuccessful, and many experts believe that reforestation may not be a significant carbon sink in the second half of this century.

Amazon rainforest clearance
Vast areas of the Amazon rainforest in Brazil have already been cleared to make way for agricultural use.

RATE OF CHANGE IN FORESTED AREA

This map shows the amount of cover and change in forested area between 2000 and 2012. Note that the highest rates of deforestation are largely in the tropics and subtropics.

0%   >80%
TREE COVER   FOREST LOSS   LITTLE CHANGE   FOREST GAIN

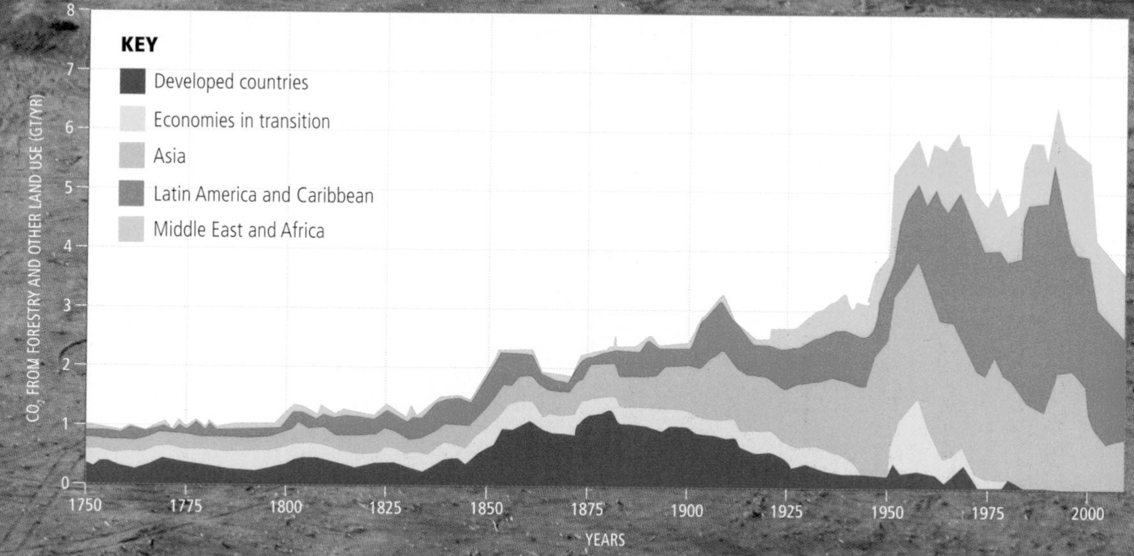

KEY

- Developed countries
- Economies in transition
- Asia
- Latin America and Caribbean
- Middle East and Africa

CO_2 FROM FORESTRY AND OTHER LAND USE (GT/YR)

YEARS

1750 1775 1800 1825 1850 1875 1900 1925 1950 1975 2000

HISTORICAL TRENDS IN FOREST CARBON EMISSIONS

This graph shows historical trends in forest carbon emissions for the period between 1750 and 2010, in Gt CO_2 eq per year. The developed countries (dark blue) have become net carbon sinks after a long history of deforestation.

Waste not, want watts?

Life pollutes: humans, cats, ferns, and bacteria all pollute. Organisms metabolize food to gain the energy they need to grow, reproduce, and move. And metabolism creates waste products; these products are often toxic and need to be eliminated. While some level of pollution is unavoidable, humans pollute for reasons other than just simple metabolism, and some of our waste products are unusable by other species, making us a unique producer of "true waste."

Landfill issues

Although waste disposal is a significant problem facing an ever-expanding and consumptive world population, greenhouse gas emissions from the waste management sector account for only about 3% of total global emissions (1.5 Gt CO_2 eq per year ◀ p.168). The largest share of waste-related emissions comes from landfills, primarily in the form of methane. The bacteria that decompose waste in the oxygen-depleted interior of landfills produce methane, but well-aerated landfills produce carbon dioxode instead of methane, which molecule-for-molecule is a weaker greenhouse gas. Of course, this methane could be used to generate energy (see below).

Researchers digging in landfills have exhumed decades-old hot dogs that showed no signs of decomposition.

Electricity from garbage
In this landfill In Livermore, California, a well collects methane generated by decomposing garbage in the subsurface. The methane can then be used to generate electricity.

In the context of the carbon cycle, those hot dogs—and much of the other waste accumulating in typical, inefficient landfills—are carbon "sinks." To some extent carbon sinks offset the release of carbon dioxide from fossil-fuel burning. For example, a corn plant may take up a carbon dioxide molecule recently emitted from a coal-fired power plant. The corn is then fed to a pig, and the pig is slaughtered to make hot dogs. A hot dog is discarded by a child at a ball game and ends up in a landfill, where it may sit for centuries without decomposing and releasing its carbon back into the atmosphere. Taken together, all the hot dogs and the rest of the waste generated globally each year "store" about 0.2 Gt CO_2 eq. This number is small, even compared to the relatively low total waste sector emissions, so we certainly don't want to produce more waste just to sequester a small amount of carbon.

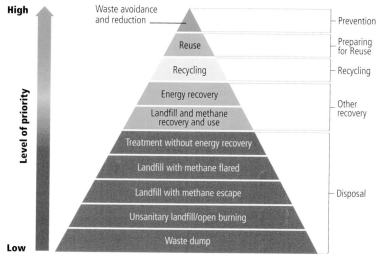

Energy from waste

Rather than attempting to further reduce already-low waste sector emissions, some communities are exploring waste recycling alternatives, including burning landfill garbage as a fossil-fuel substitute. Society is already obliged to collect waste and transport it to a central repository; why not burn this renewable energy source rather than allowing it to accumulate and release methane? State-of-the-art technology can prevent much of the potential pollution generated by incineration. The methane seeping out of landfills can also be captured

WASTE HIERARCHY

Best practices for the management of municipal waste follows a hierarchy where waste avoidance and reduction is of highest priority, followed by reuse, recycling, and energy recovery. Presently, only 20% of municipal waste is recycled and only 13.5% is treated for energy recovery, so significant gains are possible. There are significant variations from country to country in the proportion of waste that goes to landfill, from near 0% in Japan, Austria, Germany, the Netherlands, Belgium, Switzerland, and Sweden to nearly 100% in Russia, Romania, Brazil, and Turkey. Europe generates about 60% as much waste per capita as the U.S. Working our way up the waste management pyramid will have many co-benefits beyond greenhouse gas emissions, including energy recovery and less demand for landfills.

and used for energy production, further reducing the overall impact of landfills on the environment. This large-scale garbage "recycling" may turn out to be a win–win situation for society as we struggle to find ways to mitigate emissions and efficiently manage waste. In the U.S., the emissions of greenhouse gases from waste are on the decline; methane emissions from landfills declined 27% from 1990 to 2010, likely the result not only of methane capture, but of increased composting and recycling of decomposable materials like cardboard.

Geoengineering
Having our cake and eating it too

Geoengineering is an alternative approach to mitigation (◀ p.154) that involves using technology to counteract climate change impacts either at the source level (doing something about growing greenhouse gas levels) or at the impact level (offsetting climate change itself). Both approaches involve planetary-scale environmental engineering the likes of which society has never before witnessed.

Carbon sinks

One source-level geoengineering proposal, called "iron fertilization," involves adding iron to the upper ocean. Iron is a scarce nutrient in the upper ocean. This scarcity of iron limits the activity of marine plants that live near the ocean surface. Some scientists think that iron fertilization can increase the rate at which plants in the upper ocean take up CO_2, thus boosting the efficiency of the deep-ocean carbon sink (◀ p.106), and offsetting the buildup of carbon dioxide in the atmosphere. However, limited experiments suggest that iron fertilization would simply speed up cycling of carbon between the atmosphere and the upper ocean, with little or no burial of carbon in the deep ocean. And there could be negative side effects if humans interfere further with the complex and delicate ecology of the marine biosphere. Other geoengineering approaches include attempts to increase the efficiency of terrestrial carbon sinks by planting more trees and "greening" regions that are currently deserts. Many consider this approach more environmentally friendly than other schemes, but it is unclear if it could be accomplished on the scale required to significantly offset human carbon emissions.

Carbon capture

Closely related to regional greening plans are carbon capture and sequestration (CCS) approaches. In CCS approaches, carbon is extracted from fossil fuels as they are burned, preventing its escape and buildup in the atmosphere. The captured carbon is then buried and trapped beneath Earth's surface or injected into the deep ocean, where it will likely reside for many centuries. One potentially effective CCS scheme would involve scrubbing CO_2 from smokestacks and reacting it with igneous rocks to form limestone. This mimics the way that nature itself removes CO_2 from the atmosphere over geological timescales (◀ p.106). Klaus Lachner of Columbia University argued for a related alternative, in which massive arrays of artificial "trees" take carbon directly out of the air and precipitate it in a form that can be sequestered.

Saltwater in the sky
This artist's conception shows a proposed device for spraying large quantities of seawater into the atmosphere to help boost the sun-reflecting power of marine stratocumulus clouds.

Solar shields and aerosols

A frequently proposed impact-level geoengineering approach involves deliberately decreasing the amount of sunlight reaching Earth's surface so that the reduction in incoming radiation offsets any greenhouse warming. One method involves deploying vast "solar shields" in space that reflect sunlight away from Earth. Shooting sulphate aerosols into the stratosphere to mimic the cooling impact of volcanic eruptions (◀ p.18) is a less costly but potentially more dangerous alternative. This method could exacerabate the problem of ozone depletion by tampering with the chemical composition of the stratosphere.

While calculations suggest that either of these impact-level methods could offset greenhouse warming of the atmosphere, each has problems. First, they do nothing to avert the problem of ocean acidification associated with increasing atmospheric CO_2 levels (◀ p.126). Furthermore, climate models indicate that reducing the incoming solar radiation, while potentially offseting the warming of the globe, would not necessarily counteract the regional impacts of greenhouse warming. Some regions might warm at even greater rates, and patterns of rainfall and drought could be dramatically altered. And if, for some reason, these methods were ultimately halted, the full impact of warming that had been masked for decades would suddenly be unmasked, leading to dramatic, rapid global climate change.

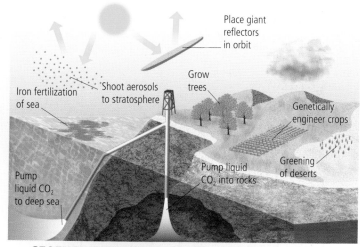

Place giant reflectors in orbit

Grow trees

Shoot aerosols to stratosphere

Iron fertilization of sea

Genetically engineer crops

Greening of deserts

Pump liquid CO_2 into rocks

Pump liquid CO_2 to deep sea

GEOENGINEERING SOLUTIONS TO CLIMATE CHANGE

The geoengineering schemes illustrated here could manipulate the composition of our atmosphere and oceans to offset the impacts of burning fossil fuels.

Schemes of last resort

Each of the proposed geoengineering schemes has possible shortcomings and poses a potential danger. Some advocates maintain that if we are backed into a corner and faced with the prospect of irreversible and dangerous climate change, we may need to resort to these schemes at least as partial solutions. Industrialists like former Microsoft CEO Bill Gates and Sir Richard Branson of the U.K. have thrown their weight and money behind geoengineering research. Some scientists advocating geoengineering have gone so far as to form start-ups that would profit from the implementation of these schemes. Others note that it would be wise not to tamper with the climate, the workings of which we still do not entirely understand. Either way, the debate over whether geoengineering is likely to be an effective and prudent solution to climate change is bound to continue—as scientists continue to propose new technology to address climate change problems.

But what can I do about it?

If all this talk about energy sectors and governmental buy-in leaves you feeling helpless in the face of global warming, don't let it! Our lifestyle choices can directly aid in the mitigation of greenhouse gas emissions. Often, these are "no regrets" changes that have positive side benefits—improving our quality of life, conserving natural resources, and facilitating greater environmental sustainability.

Lifestyle choices

First, we can be more efficient in our use of energy. We can make home improvements that decrease the energy we use to heat and cool our houses and apartments. More efficient practices include better insulation, passive solar heating, and using fans or opening windows for air conditioning. We can replace inefficient incandescent light bulbs with more efficient bulbs. An important recent trend involves "smart houses," where thermostats can be programmed and remotely controlled for maximum efficiency, where occupancy sensors can be used to minimize the unnecessary use of lighting, and smart power strips eliminate "phantom energy" by automatically sensing power use and reducing power drainage by appliances in standby mode. There is also significant mitigation opportunity in simply being better about recycling.

There are other changes we can make that don't require that we remodel or even buy new appliances. Clotheslines make an excellent substitute for dryers, and unplugging appliances that are not in use helps reduce electricity leakage.

We can make serious contributions to emission reduction efforts with our transportation choices. Many of us could commute to work by bicycle or on foot. For those of us who have difficulty finding

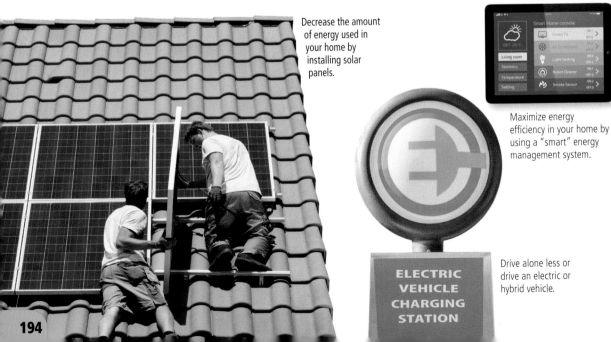

Decrease the amount of energy used in your home by installing solar panels.

Maximize energy efficiency in your home by using a "smart" energy management system.

ELECTRIC VEHICLE CHARGING STATION

Drive alone less or drive an electric or hybrid vehicle.

Clotheslines make an excellent substitute for electric dryers.

time to maintain fitness regimes, this option allows for the best sort of multitasking—we exercise while reducing our carbon footprints.

Other alternatives include public transportation and carpools. Hybrid and electric vehicles are another exciting new option. Given the high cost of gasoline in recent years, this option not only benefits the environment, but our pocketbooks as well.

Education and incentives

Employers, governments, and non-governmental organizations can play an important role. Community-focused organizations can provide relevant guidance and education to individuals. Some governments already provide tax benefits and incentives for citizens who build green, add solar panels to their roof, or buy hybrid vehicles. Public outreach efforts can also include educational programs that teach energy conservation practices, and campaigns aimed at encouraging individuals to make environmentally conscious decisions. If you want to know how well you are doing in terms of your own contribution to global greenhouse gas emissions, turn to p.198.

Commute to work by bicycle or on foot.

Replace incandescent bulbs with energy-efficient bulbs.

Appliances that are not in use can be unplugged, reducing electricity leakage.

Remember to recycle.

Sustainability success stories

As the authors of this book well know, the challenges posed by anthropogenic climate change can seem overwhelming, even depressing, especially to students in our college classrooms. But there are uplifting success stories that show how steps can be taken to reduce carbon footprints and create a more sustainable future for generations to come. And we don't have to look too far—outside our offices, banks of recycling bins have been distributed on every floor of buildings all over campus, aimed at the goal of zero landfill.

The program at Penn State is called "Mobius," recycling with a twist: a goal of zero landfill, with a closed recycling loop that has no beginning and no end. The institution currently diverts 69% of its solid waste from landfill, and with a recent new initiative in composting, will reach 75%. And adding miscellaneous plastics to the recycling program will prevent 85% of its solid waste from reaching landfills.

Becoming more efficient

Of course, energy use is the largest source of emissions for a campus. Penn State, like many other institutions and corporations, has committed to reducing energy consumption through conservation and increased efficiency. Since 2005, Penn State has reduced its campus greenhouse gas emissions by 18%, and has set a goal of 35% reduction by 2020. Toward that end, the campus steam plant is converting from coal to natural gas, with the expectation of significant reductions in emissions intensity (carbon emitted per unit of energy generated). Their expected goal may be fulfilled, but the full "life cycle" assessment of climate and other environmental impacts of various energy alternatives remains an area of active research.

Penn State's fleet operations unit is converting to high fuel-efficiency vehicles and encouraging the use of public transportation. Campus food services use locally grown food where available. And energy conservation has become policy designed to lower energy consumption through employee and student action. Occupants of campus buildings play a major part in reducing energy use because about 30% of energy use is under the control of building occupants. Employees and students are expected to turn lights off in unoccupied spaces; maintain comfortable but moderate temperatures in offices, dormitories, and classrooms; turn off

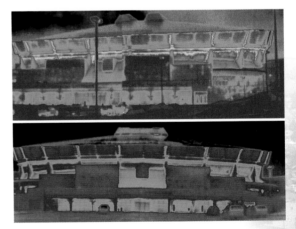

Cutting heat loss
These before (above) and after (below) infrared photos of the Bryce Jordan Center (right), a basketball arena and concert venue on the Penn State campus, show how better insulation and closing the undersides of overhanging structures (soffits) drastically reduced building heat loss as indicated by the elimination of the yellow and red colors in the lower photo.

Recycling waste
Penn State minimizes landfill by providing a variety
of recycling bins around the campus.

College students of today will provide the solutions to sustainability challenges of the future. Green think tanks are appearing on campuses like Harvard University, where their "Green Team" harnesses the power of student creativity and interest in environmental issues to help Harvard achieve its sustainability goals. Staff and community efforts round out the sustainability initiatives at this and many other campuses. Major corporations are partnering with organizations like the Student Conservation Association to provide sustainability internships to college students. These partnerships develop business skills as well as awareness of environmental sustainability challenges faced by the private sector and companies, and the methods they use to address these challenges. The enthusiasm, creativity, and entrepreneurial spirit of young people is the antidepressant we take every day as we confront the challenges posed by the often wasteful use of planetary resources by an increasingly consumptive human society.

computer monitors when not in use; use shared printers, refrigerators, and other appliances; and in all cases use only appliances with the Energy Star label.

Working for the future

Sustainability has been defined as the simultaneous pursuit of human health and happiness, environmental quality, and economic well-being for current and future generations. As a guiding principle, sustainability satisfies many of the objectives we strive for in our daily lives and in the strategic goals of our institutions, corporations, agencies, and organizations.

What's your carbon footprint?

Before you go on a diet, you might weigh yourself to establish a starting point and also, perhaps, to further motivate yourself to cut back on calories. In a similar way, you can calculate your "carbon footprint"—your personal contribution to the problem of global warming—as a first step in reducing the emissions your lifestyle generates.

There are numerous carbon footprint calculators on the Internet; you may wish to try a couple to compare results. We give the U.S. EPA's URL below right. All of these calculators help you to evaluate your lifestyle and determine your personal contribution or "footprint." Carbon footprints are usually measured in metric tons of CO_2 equivalent (eq) per year.

To assess your footprint, you will be asked to answer a series of questions about your lifestyle.

Footprint questions

- Where do you live? There are regional differences in the energy sources used to generate electrical power, and these affect emissions.

- How many people live in your home? Your carbon footprint is smaller if the energy you use is shared.

- What type of vehicle and how many miles per year do you drive? Your car's gas mileage affects emissions.

- How often do you fly, and are your trips short or long? Airline travel is a big emissions contributor. Short trips use more energy per mile because takeoffs are particularly fuel-intensive.

- Do you heat your home with natural gas, heating oil, or propane? How much is your typical monthly heating bill? What is your typical monthly electric bill?

- Do you eat red meat, just chicken and fish, or are you a vegetarian? The various activities that put these foods on your table have differing carbon emission profiles.

- Do you eat mostly local foods and buy mostly local products, or do you prefer imported goods? Long-haul freight consumes fossil fuel aggressively.

- What types of recreation do you prefer? A bicycling trip that begins at your own front door contrasts quite markedly with snowmobiling or motorboating at distant recreation sites.

Now take off your shoes and compare your print to two famous footprints!

SASQUATCH

Carbon footprint:

30 METRIC TONS CO₂ EQ PER YEAR

At home
Lives in a poorly insulated, air-conditioned apartment in the American Midwest with electric heat; loves to soak in a hot bathtub several times a week; does not recycle.

On the go
Drives an SUV 24,000 km (15,000 miles) per year; takes 5–10 business trips by plane each year, both short haul and long haul; and takes one exotic personal vacation each year by plane.

CINDERELLA

Carbon footprint:

4 METRIC TONS CO₂ EQ PER YEAR

At home
Lives with two roommates in France in an energy-efficient, well-insulated apartment with electric heat; prefers taking showers; recycles a significant fraction of household waste.

On the go
Drives a hybrid car 16,000 km (10,000 miles) per year; journeys by train 1,600 km (1,000 miles) per year; uses videoconferencing rather than traveling for business; and takes one exotic personal vacation each year by plane.

Does your footprint looks more like Cinderella's dainty glass slipper or Sasquatch's big paw? Get online, find a carbon footprint calculator, and let the site suggest ways in which you can become more carbon stingy and environmentally friendly.

Carbon Footprint Calculator

http://goo.gl/zQIoxN

Global problems require international cooperation

The atmosphere does not recognize national boundaries. When pollutants such as greenhouse gases or industrial aerosol particulates are emitted, they travel great distances, crossing continents and oceans. No single nation can solve the problems created by atmospheric pollutants. Some argue that, because of its pervasiveness, global warming is so daunting that humans cannot solve a problem so huge in scale—especially one that requires cooperation between so many disparate parties. However, history suggests otherwise, and even provides precedent for cooperation between nations in solving environmental problems.

The effects of acid rain
This conifer forest on Mount Mitchell, in North Carolina, includes many trees damaged or killed by acid rain.

Acid rain

By the 1970s, large parts of the northeastern U.S. and eastern Canada were plagued by the problem of **acid rain**. As waterways became acidified, fish populations and other aquatic life died off in startling numbers. Acidic rainfall began to kill trees as well. Research implicated the sulphate and nitrate aerosol particulates produced by factories located in the American Midwest. Aerosols were being carried downwind, where they dissolved in rainfall to form sulfuric and nitric acid. Europe experienced similar acid rain problems. In some cases, acid rain was destroying historic monuments and structures, as well as natural habitats.

"The Montreal Protocol to ban ozone-depleting chemicals was perhaps the single most successful international agreement to date."

Former U.N. Secretary General Kofi Annan

In response, the U.S. passed a series of laws, culminating with the Clean Air Act of 1990. This act included specific provisions for dealing with the acid rain problem. The Clean Air Act of 1990 (and other clean air acts passed in the U.S. and other nations) led to the widespread introduction of "scrubbers" into factories, which remove harmful particulates from industrial emissions before they enter the atmosphere.

Ozone depletion

Perhaps an even better analogy to the challenge posed by climate change is the problem of ozone depletion. A breakdown in the stratospheric ozone layer was measured at the South Pole in the early 1980s. Theoretical and observational considerations pointed to an anthropogenic cause. Industrial products such as chlorofluorocarbons (CFCs), used at the time as refrigerants and propellants in aerosol cans, were eventually reaching the stratosphere. There, in the presence of solar radiation, they produced chemicals capable of destroying the ozone layer.

Since the ozone layer prevents most harmful ultraviolet radiation from reaching Earth's surface, the depletion of this protective layer was tied to an increase in skin cancers and other damaging effects on plants and animals. In 1989, worldwide concern led to adoption of the Montreal Protocol, an international agreement banning the production of ozone-depleting chemicals.

Dying rivers
Acid rain has a devastating effect on aquatic environments. This sign in Nova Scotia gives some protection to a river's already depleted salmon stocks.

CFCs
CFCs were used as propellants in spray cans until they were implicated in the destruction of the ozone layer.

A global challenge

Arguably, the problem of climate change is more challenging to solve than that of ozone depletion. In that case, other commercial refrigerants and propellants were readily available as substitutes for ozone-depleting substances. By contrast, the emission of greenhouse gases results from the world's dependence on fossil fuels—its primary source of energy—and unfortunately a substitute for fossil fuels that can meet current (and future) world energy demands has yet to be found. Emission control and reduction requires a fundamental change in global energy policies.

The Kyoto Protocol

Efforts to achieve just that include:

- **1992:** The United Nations Framework Convention on Climate Change (UNFCCC), an international treaty, was formulated at the "Earth Summit" held in Rio De Janeiro, Brazil.

- **1997:** Five years later, an update to the UNFCCC—the "Kyoto Protocol"—was agreed upon at a summit in Japan.

- **2005:** Only after another eight years did the treaty actually go into force. The stated objective of the Kyoto Protocol is to achieve "stabilization of greenhouse gas concentrations in the atmosphere at a level that would prevent dangerous anthropogenic interference with the climate system."

(Cont.)

Logging concessions
A truck hauls freshly cut logs in the Limbang area of Sarawak, Borneo. Once covered with pristine primary forest, the area has been devastated by logging and is also under threat from dam projects that will flood large areas of forest, destroying unique ecosystems and forcing local indigenous groups from their ancestral homelands. Despite projects aimed at conserving tropical forests, which would preserve ecosystems as well as mitigate the increase in atmospheric CO_2, deforestation remains a major problem.

The Kyoto Protocol officially expired in 2012—192 parties ultimately ratified the treaty, thereby committing to reducing greenhouse gas emissions to mandated levels. The two industrialized nations that have not ratified the protocol are the U.S. and Canada, the latter having actually ratified the treaty before backing out later under conservative Prime Minister Stephen Harper. A large number of developing nations also ratified the protocol, but they are not held to mandated reductions due to the financial hardships the reductions might impose upon their fragile economies.

The protocol included provisions to ensure that the developed world assists developing nations in moving toward more environmentally friendly energy resources.

The Kyoto Protocol has been criticized by both sides in the climate change debate. Critics on one side argue that it didn't go far enough, and that the emission cuts mandated in the protocol would not stabilize greenhouse gas concentrations below dangerous levels. Supporters point out, however, that it was just a first step,

putting in place a framework that can be built upon in the future to achieve broader reductions. Critics on the other side argue that committing to Kyoto would destroy the global economy. Yet cost-benefit analyses suggest that the cost of inaction could be far greater (◀ p.156). This debate is arguably more about politics than objective scientific or economic considerations and is likely to continue into the foreseeable future.

Post-Kyoto period

A series of recent negotiations have sought to establish a new international framework to replace and build upon the now-expired Kyoto Protocol. In 2007, under the Bali Action Plan, all the developed countries agreed to a "roadmap" involving "quantified emission limitation and reduction objectives, while ensuring the comparability of efforts among them, taking into account differences in their national circumstances."

As part of the December 2009 Copenhagen summit, countries agreed to the Copenhagen Accord, stating that global warming should be limited to below 2.0°C (3.6°F) warming, understood to be measured relative to pre-industrial temperatures. Many observers expressed frustration at the lack of any binding agreement over greenhouse gas reductions. But the accord did result in major emission-reducing commitments from countries such as China and India, and more pledged emissions reduction than the Kyoto Protocol achieved. The EU and 17 developed nations have submitted specific carbon reduction targets as part of the Accord.

Some features of the Copenhagen Accord were incorporated into the UNFCCC process at the 2010 summit in Cancún, Mexico, including adoption of the 2.0°C (3.6°F) warming target and recommendation of a potentially even more stringent 1.5°C (2.7°F) target "on the basis of the best available scientific knowledge" regarding potential climate change risks.

Participating nations submitted carbon reduction plans to the UNFCCC in Cancún, supplementing commitments already made as part of the Bali Action Plan in 2007.

At the Durban, South Africa, summit of 2011, all parties agreed to develop a new emissions reductions agreement with legal force under the UNFCCC. At the Doha, Qatar, summit of 2012, nations launched a new commitment to reach universal climate agreement by 2015. Because no binding reductions agreement has yet been reached, there is concern that the opportunity to limit warming below 2.0°C (3.6°F) is quickly slipping away. One obstacle remains the failure of the industrial world's leading emitters to reach internal agreement with respect to energy and climate policy.

Because the U.S. Congress failed to pass a comprehensive climate bill, many states and localities have implemented their own greenhouse gas reduction schemes. A coalition of west coast states including California, Oregon, and Washington and British Columbia are developing their own cap-and-trade system, as is a coalition of northeastern and mid-Atlantic states.

In the absence of congressional action, the Obama administration has pursued executive actions via the EPA to reduce U.S. carbon emissions. Measures include imposing tighter restrictions on carbon emissions from coal-fired power plants and passing higher automobile fuel-efficiency standards. These actions are aimed at meeting a target of 17% reductions relative to 2005 carbon emission levels by 2030.

The administration has signaled it will pursue a binding international treaty on emissions reductions with or without ratification from a currently intransigent U.S. Congress at the December 2015 U.N. summit in Paris.

Commitment to reduce emissions
U.N. Secretary General Ban Ki-moon addresses delegates at the 20th session of the Conference of the Parties on Climate Change (COP20) and the 10th session of the Conference of the Parties serving as the Meeting of the Parties to the Kyoto Protocol (CMP10) in Lima, Peru, December 2014.

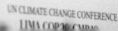

Can we achieve sustainable development?

A responsible society strives to meet its needs without compromising the ability of future generations to meet theirs. This defines sustainable development. Sustainability requires that we protect ecosystems from destruction and consume natural resources at a rate no greater than nature can provide. Achieving environmental sustainability is difficult in terms of water use and soil erosion, and seemingly impossible when we take into account fossil-fuel and mineral-resource consumption. Since cutting consumption is such a challenge, we should also look to renewable energy and recycling to achieve sustainability goals. Nevertheless, considerable progress has been made in improving energy efficiency with recent development of appliance and building energy-efficiency standards and labeling. On top of that, significant growth in renewable energy has occurred in the last several years, led by widespread and expanded electricity-generating capacity of wind, hydro, and solar power, promoted by favorable energy policies involving tariffs and quotas.

Developing nations

Issues of equity enter into the sustainability equation because current models of economic development depend on increased consumption and depletion of natural resources. The challenge is to find ways that developing countries can achieve a quality of life equal to that of the developed world without damaging the environment and depleting resources. Doing so requires that they make a substantial shift away from the highly consumptive and largely unsustainable path followed by the developed world. Fortunately, developing nations can now utilize previously unavailable technologies that may help them to meet their needs with reduced impact on the environment.

For example, China has slowed its fossil-fuel-use increases with a combination of activities. By shifting to renewable and less carbon-intensive energy sources, imposing economic reforms, and slowing population growth they have moved in a positive direction. In September 2014 they launched

DEVELOPMENT STRATEGIES

There are many strategies for mitigating against climate change. Most of these enhance sustainability but also involve trade-offs.

	Mitigation option		
	Improving energy efficiency	Reforestation	Deforestation avoidance
Compatibility with sustainable development	Cost effective; creates jobs; benefits human health and comfort; provides energy security	Slows soil erosion and water runoff	Sustains biodiversity and ecosystem function; creates potential for ecotourism
Trade-offs		Reduces land for agriculture	May result in loss of forest exploitation income and shift to wood substitutes that produce more emissions

a nationwide cap-and-trade program intended to put a price on carbon that should cut CO_2 emissions to 40–45 percent of the 2005 levels by 2050. India, Turkey, Mexico, South Africa, Russia, and Brazil are also working to decouple economic development from fossil-fuel dependency.

Strategies for sustainable development, including expansion of the use of alternative energy and carbon sequestration, have clear benefits for society (◄ p.192). These technologies can enhance national security by reducing dependency on foreign oil, creating new jobs, and stimulating economies (see table below). But in some instances, short-term economic benefits may conflict with environmental benefits. For example, the shift in developing countries from biomass (wood-fire) cooking to the use of cleaner and more efficient liquid propane (fossil-fuel) stoves enhances health and quality of life by reducing indoor pollution, but increases dependency on fossil fuels. This, in turn, increases greenhouse gas emissions and exacerbates human-induced climate change.

Developing policy

The responsibility for implementing sustainable development policies lies with government, industry, and civil society:

■ Just as human health has improved globally thanks to diverse local strategies, progress toward a common goal such as climate change mitigation can be made via disparate governmental policies.

■ Sustainability and profitability are gradually being seen as compatible goals in industry, perhaps essentially so for large, multinational companies. Regulatory compliance is also a factor, but may not be as important as was once thought.

■ Non-governmental organizations (NGOs), which encourage reform, provide policy research and advice, and champion environmental issues, increasingly express the will of civil society. Academia also plays a role, especially in research, which enhances understanding of the scientific, economic, and political implications of climate change. The IPCC itself is a prime example of how a civil body can influence the world's response to global warming.

Fortunately, the goals of climate change mitigation and sustainable development are largely compatible. Together, these two strategies can help us to create a healthier and more durable society for the future.

Incineration of waste	Recycling	Switching from domestic fossil fuel to imported alternative energy	Switching from imported fossil fuels to domestic alternative energy
Energy is obtained from waste	Reduces need for raw materials; creates local jobs	Reduces local pollution; provides economic benefits for energy-exporters	Creates new local industries and employment; reduces emissions of pollutants; provides energy security
Air pollution prevention may be costly	May result in health concerns for those employed in waste recycling	Reduces energy security; worsens balance of trade for importers	Alternative energy sources can cause environmental damage and social disruption, e.g., hydroelectric dam construction

The ethics of climate change

The international media has paid considerable attention to the economic implications of global climate change. By contrast, they have paid little attention to the equally important ethical considerations. The objective of the Kyoto Protocol—to stabilize "greenhouse gas concentrations...at a level that would prevent dangerous anthropogenic interference with the climate system"—begs several questions. Who, for example, determines what constitutes "dangerous"? Answering such questions requires us to take into account political, cultural, and philosophical principles that are fundamentally ethical in nature.

Winners and losers

One tricky ethical principle is "equity." Equity issues surrounding climate change include the fair distribution of risks, benefits, responsibilities, and costs to both developed and developing nations. Climate change will be associated with potentially dramatic redistributions of wealth and resources, impacting food production, fresh water availability, and environmental health. In the course of this shuffling, there will be winners and losers. Unfortunately, climate change won't play fair. In fact, climate change may play the role of a "reverse Robin Hood," taking resources from the poor and giving them to the rich. Tropical regions—essentially the developing world—will likely suffer the most detrimental impacts. In the short term, the developed nations may even stand to benefit. European and North American agribusinesses, for example, may enjoy longer growing seasons (◀ p.140).

Further ethical complications arise from the fact that the individuals who gain from current fossil-fuel burning are not the same as the individuals who stand to lose when the climate changes. Is it possible to assign a meaningful cost to the devastating impacts of climate change on the poor and disadvantaged?

After the flood
A Bangladeshi woman collects water from a well submerged by floodwater. About three-quarters of the country is less than 10 m (33 ft) above sea level and flooding is common. However, rising sea levels increase the risk of severe flooding and the resulting loss of life, property, and infrastructure.

What is the value of the life of a starving child in Bangladesh as measured in cheap barrels of oil? Do we even dare to pose such questions?

And the developing world, by virtue of its relative poverty and lack of technological infrastructure, is far more vulnerable to the economic, environmental, and health threats posed by climate change. Ethical considerations would seem to demand that the developed world assist developing nations in adapting to climate change, both in mitigating impacts and exploiting possible benefits.

The developed world has already benefited from a century of cheap fossil-fuel energy. Is it fair to tell developing nations, who are just now beginning to build their energy and transportation infrastructures, that they can't have their turn to enjoy cheap oil? This challenging ethical dilemma complicates discussions about the appropriate burden of mitigation, including the distribution of emissions rights both among nations and between generations.

Social discounting

When it comes to the generational transfer of the benefits and costs of fossil-fuel burning, "social discounting" places a greater value on benefits today at the expense of subsequent generations. This is based on the assumption that future generations will have access to new technology and will be better equipped to deal with environmental challenges. Social discounting involves making an ethical call. If we discount future potential impacts too strongly and assume that future generations will be able to solve all problems, the cost-benefit analysis will surely favor inaction. Is it fair to gamble like this, knowing that it is our grandchildren who will pay the price if our assumptions turn out to be wrong?

Geoengineering dilemmas

Ethical considerations are also raised by geoengineering approaches to mitigation (◀ p.192). Should nations that stand to benefit from certain types of intervention be able to do so, even when other nations may be negatively impacted by their actions? Ethical issues complicate discussions of biofuel technology, too. Should agricultural land currently used to feed people be reallocated for energy production at a time when starvation and malnourishment are omnipresent?

An essential step in tackling these problems is for all countries—including holdouts like the United States—to join the international effort to stem the buildup of greenhouse gases initiated by the Kyoto Protocol (◀ p.201).

The known unknowns & unknown unknowns

There are at least two kinds of unknowns. There are the "known unknowns," which are the questions we already know to ask, but for which we don't yet have the answers. Then there are the "unknown unknowns." These are the questions we don't even know to ask, the questions involving phenomena that currently lie beyond the horizons of our imagination.

A great deal of discussion in this book is devoted to the known unknowns. We have discussed the open scientific questions regarding how much warming is to be expected and precisely what the pattern of climate change will be. These uncertainties are linked to unknowns regarding the societal and environmental impacts of climate change (e.g., changes in water availability, food supply, and disease prevalence). We have examined the still-unsolved mysteries of the great climate changes in Earth's past, and the changes in violent weather phenomena, such as hurricanes, that may lie in store for us in the future.

More known unknowns

The known unknowns also include the lack of certainty regarding the "tipping points" looming in our future. Scientists recognize that such tipping points probably exist, but they don't know exactly where they may lie:

- Just how rapidly will the major ice sheets melt, and how high will the sea level rise accordingly? (Recent studies indicate that the loss of ice from the Antarctic and

"Drunken" trees
These treees in Fairbanks, Alaska, have fallen due to melting of the permafrost beneath them, itself a result of rising Arctic temperatures. Permafrost melting also results in the release of the greenhouse gases methane and carbon dioxide, although the full effects of this release on the climate are not yet known.

Greenland ice sheets (◀pp.110–111), and associated contributors to sea level rise, may be proceeding faster than concluded even in the latest IPCC report.)

- Will the "conveyor belt" ocean circulation weaken? And if so, when?

- Will the ability of the oceans and plants to absorb the CO_2 we are adding to the atmosphere change in the future?

Also included in the known unknown category are answers to questions relating to the unpredictability of human behavior.

- What will future human-driven emissions patterns be?

- What will the economic implications of warming be?

- What steps will we take to mitigate against greenhouse gas buildup and climate change? How successful will mitigation efforts be?

- Will we implement any of the currently conceived geoengineering plans? Will new, risk-free plans be conceived of?

Unknown unknowns

And what about the unknown unknowns? There are some of these in the science itself:

- Will the response of the climate to increased greenhouse gas concentrations take an unpredicted course?

- What are the tipping points that have not been conceived of yet?

Into the unknown
Like this scuba diver exploring an underwater cave in Mexico's Yucatan Peninsula, we might discover previously unsuspected phenomena as we continue to investigate climate change.

- Are there hidden reserves of carbon on our planet that could suddenly be released, leading to further warming? (Recent observations of methane escaping from Arctic permafrost and continental shelves suggest the possibility that certain feared carbon cycle feedbacks may already be kicking in.)

In the case of adaptation and mitigation, the unknown unknowns may be the stuff of science fiction. Decades ago, who would have imagined modern-day technology such as cloning or hand-held "smart phones" as powerful as the supercomputers of previous decades? More to the point, who would have conceived of modern transportation options such as hybrid vehicles, or prospective energy technology such as creating genetically altered bacteria that generate propane rather than produce cell membranes?

So what are we to make of all of this uncertainty?

Clearly, we must work to diminish the uncertainty where possible, particularly when it impacts our ability to make appropriate policy decisions or choose an optimal strategy for mitigating climate change. Recent history has taught us that uncertainties are not adequate justification for avoiding action. We know enough today to understand how vital it is that we act now.

The urgency of climate change
Why we must act now

Uncertainty abounds (◀ p.208) but it is a poor excuse for inaction. In fact, given the possibility of severe and irreversible harm to society and the environment, scientists generally advocate that we abide by a "precautionary principle" that puts the onus of proof on those advocating inaction.

No excuses

If the remaining uncertainty in the science is not a valid argument against taking immediate action to slow climate change, then what, if anything, is? As we have seen, some argue that action could harm the economy. Yet this argument does not appear to withstand scrutiny, since the economic harm of inaction looks to be greater (◀ p.156) in the long term. Others argue that climate change might be beneficial to humankind, but an impartial assessment strongly suggests otherwise (◀ p.116). Still others concede that climate change represents a potential threat, but that it is only one of many problems facing society, and that focusing on climate change issues might divert attention and resources from more pressing problems. The argument that we must choose between competing societal problems, however, is based on the flawed premises that society can only solve one problem at any given time, and that the problems facing society are independent of each other. In the case of climate change, we have already shown its potential to exacerbate other major global societal and environmental issues, including:

- Sustainability
- Regional conflict
- Biodiversity
- Extreme weather events
- Water availability
- Disease

Some proponents of inaction argue that we can engineer our way out of the problem with future technological "fixes." However, the potential pitfalls of high-tech fixes present risks as well. And many of these fixes may not be able to prevent or reverse the more serious consequences of climate change, such as the melting of the Greenland ice sheet (◀ p.192). While there are some promising new technologies on the horizon, there are none currently available that will handily satisfy our global thirst for carbon-free or carbon-neutral energy in the decades to come. Furthermore, while carbon sequestration (CCS) may ultimately slow the buildup of carbon dioxide in the atmosphere, the feasibility of large-scale implementation of CCS has yet to be demonstrated (◄ p.192).

Our children's and grandchildren's world

If we choose not to act on this problem now, then in the very best-case scenario we must accept that our children and grandchildren will grow up in a world lacking some of the beauty and wonder of our world. They may come of age in a time where:

- Polar bears, pikas, and numerous other creatures will be the stuff of myth

- There will be no Great Barrier Reef to explore

- Giraffes and elephants will no longer loom in the foreground of the majestic snows of Kilimanjaro

- Great coastal communities such as Amsterdam, Venice, Miami, and New Orleans will disappear beneath the waves

Of course, humankind might plausibly adapt to these sad changes.

In the worst case scenario, however, our grandchildren will grow up, as renowned climate scientist James Hansen has bluntly put it, on "a different planet"— a planet potentially resembling the dystopian world depicted in science fiction movies such as *Soylent Green* and *The Island*. Adaptation, in this case, is unlikely to be viable for many of the world's people and other living things.

Climate change has been described as a problem with a huge "procrastination penalty." With each passing year of inaction, stabilizing Earth's climate becomes increasingly difficult.

New face of climate change
Walruses usually spend summer months on the fringes of Arctic sea ice, but the disappearance of the ice may have dire consequences for the species.

Our greatest challenge

There is no "silver bullet" that will solve the problem of global climate change. But that does not mean we should throw up our hands in the face of this urgent problem. Any viable solution is going to require action from many governments and all strata of society; it will involve adapting to the changes that are inevitable, and mitigating the changes we can avert. It goes without saying that alternative energy sources must be aggressively developed and deployed, and that governments must incentivize and reward responsible behavior by individuals and corporations.

The future in our hands
Our planet has supported life for billions of years, but only over the past century has a species—humans—developed the ability to alter the planetary environment. Will we do good or harm with this newfound ability? The answer is in our hands.

Climate change is the greatest challenge ever faced by human society. But it is **a challenge that we must confront**, for the alternative is a future that is unpalatable, and potentially unlivable. While it is quite clear that inaction will have dire consequences, it is likewise certain that a concerted effort on the part of humanity to act in its own best interests has great potential to result in success.

Glossary

The first mention of each of these glossary terms is highlighted by **bold type** in the text.

Acid rain

Acid rain refers to any form of precipitation (e.g., rain, snow, sleet) that is unusually acidic. Largely caused by industrial emissions of sulfur and nitrogen aerosols, which form sulfuric and nitric acid when combined with water droplets suspended in the atmosphere, acid rain has caused documented damage to trees and plants, fish and other aquatic animals, building facades, and monuments.

Aerosols

Aerosols are microscopic liquid droplets, dust, or particulate matter that are airborne in the atmosphere. An aerosol may remain suspended in the atmosphere for hours or years depending on the type of aerosol and its location in the atmosphere. Aerosols can be of either human or natural origin and they may reflect and/or absorb incoming and outgoing radiation. This scientific definition of aerosol should not be confused with so-called "aerosol" spray cans and their contents. (See glossary entry for Chlorofluorocarbons.)

Anthropocene

The most recent epoch of Earth history, during which human activity has notably impacted Earth's climate, surface features, and biological diversity.

Atmosphere

The atmosphere is the gaseous envelope surrounding Earth, which is retained by Earth's gravitational pull. The first 80 km (50 miles) above Earth contains 99% of the total mass of Earth's atmosphere and is generally of a uniform composition (except for a high concentration of ozone, known as the stratospheric ozone layer, at 19 to 50 km/12 to 31 miles). The gases that make up the atmosphere are nitrogen, 78%; oxygen, 21%; argon, 1%; carbon dioxide, 0.04%; and minute traces of neon, helium, methane, krypton, hydrogen, xenon, and ozone as well as trace amounts of water vapor, the distribution of which is highly variable. Earth's atmosphere features distinct layers: the troposphere, the stratosphere, the mesosphere, the thermosphere, and the exosphere. The term "free atmosphere" refers to the portions of the atmosphere that lie above the troposphere. (See glossary entries for Stratosphere and Troposphere.)

Biofuels

Biofuels are solid, liquid, or gaseous fuels consisting of, or derived from, biomass (plant material, vegetation, or agricultural waste used as energy sources). Biofuels can aid in the mitigation of greenhouse gas emissions by providing carbon neutral alternatives to fossil fuel burning. Burning biofuels does not lead to any net increase in greenhouse gas concentrations because the carbon released when biofuels burn was just recently removed from the atmosphere and has only been stored in plants for a few months. For every gram of carbon dioxide released when a biofuel is burned, a gram was removed from the atmosphere by photosynthesis just a short time ago. This balance is why biofuels are considered "carbon neutral." Liquid or gaseous biofuels can be used for transport, while solid biofuels can be burned for heat or to generate electric power. Common currently used biofuels, such as ethanol, are derived from maize (corn). Cleaner and more efficient biofuels, such as cellulosic ethanol, derived from switchgrass, are currently under development.

Biosphere

The biosphere includes all life on Earth and the physical environment that supports it. In this sense, the biosphere encompasses the parts of the outermost solid earth, soils, water- and ice-covered regions, and atmosphere that contain living organisms.

Carbon credits

Carbon credits are permits or certificates granting a government or company the right to emit a certain amount—typically 1 metric ton (about 1.1 U.S. tons)—of carbon dioxide. A limited number of such credits are issued to a given entity for reducing carbon emissions. Carbon credits and forms of taxation are mechanisms that society might use to meet objectives from treaties for stabilizing atmospheric carbon dioxide levels or avoid exceeding global average temperature thresholds.

Carbon cycle

The carbon cycle is the sum of the processes, including photosynthesis, decomposition, respiration, weathering, and sedimentation, by which carbon cycles between its major reservoirs: the atmosphere, oceans, living organisms, sediments, and rocks. (See glossary entry for Photosynthesis.)

Carbon dioxide (CO_2) equivalent

Carbon dioxide equivalent expresses the amount of CO_2 that would have the same global warming potential as a given greenhouse gas measured over some defined time frame, commonly one century. It is typically measured in gigatons ("Gt CO_2 eq"). (See glossary entry for Gigaton.)

Carbon footprint

One's carbon footprint is the total amount of carbon emissions one is responsible for through his or her day-to-day activities. Carbon footprints are typically expressed as metric tons of carbon dioxide (or its equivalent) emission per year.

Chlorofluorocarbons (CFCs)

Chlorofluorocarbons (CFCs) are synthetic compounds consisting of a carbon atom surrounded by some combination of chlorine and fluorine atoms. CFCs are powerful greenhouse gases. However, they are better known for their role in ozone depletion. CFCs were formerly widely used as refrigerants, propellants (in so-called "aerosol" cans), and as cleaning solvents. Implicated in the destruction of the protective stratospheric ozone layer in 1989, the industrial use of CFCs, and other related compounds, was prohibited by the Montreal Protocol. (See glossary entry for Ozone.)

Climate proxy

Climate proxies are indirect sources of climate information from natural archives such as tree rings, ice cores, corals, cave deposits, lake and ocean sediments, tree pollen, and historical records. Information from climate proxies can be used to reconstruct climate for times prior to the establishment of a widespread instrumental atmospheric and oceanic data set. (See glossary entry for Instrumental record.)

Cryosphere

The cryosphere is a term for the cold regions of the planet where water persists in its frozen form, i.e., regions covered with glaciers and ice sheets, or with permanently frozen soils. The cryosphere plays different roles within the climate system. The two continental ice sheets of Antarctica and Greenland actively influence the global climate and may also have effects on sea level. Snow and sea ice, with their large areas and relatively small volumes, are connected to key interactions and feedbacks on global scales, including solar reflectivity and ocean circulation. Perennially frozen ground (permafrost) influences soil water content and vegetation over vast regions, and is one of the components of the cryosphere that is most sensitive to atmospheric warming trends.

Deep time

Distant geologic history, generally before the recent ice-age cycles began 2 million years ago, is often referred to as deep time.

Ecosystem

An ecosystem consists of interdependent communities of plants, animals, and microscopic organisms, and their physical environment. All these different elements interact and form a complex whole, with properties that are unique to that particular combination of living and non-living elements. Ecosystem boundaries are generally delineated by climate: desert ecosystems in the subtropics, tropical rainforest ecosystems near the equator, and tundra ecosystems near the poles. As climates have changed in the geologic past, ecosystems have shifted in response.

El Niño

El Niño is a climate event in the tropical Pacific ocean and atmosphere wherein the trade winds in the eastern and central tropical Pacific are weaker than usual, there is less upwelling of cold subsurface ocean water in the eastern Pacific, and relatively warm water spreads out over much of the tropical Pacific ocean surface. During an El Niño event, the warmer tropical Pacific surface ocean waters influence the overlying atmosphere and alter the patterns of the extratropical jet streams of the northern and southern hemisphere and the general circulation of the atmosphere. The altered circulation of the atmosphere leads to changes in temperature and precipitation patterns in many regions across the globe. The name "El Niño" (literally "the boy child") derives from the Spanish term for the Christ Child and originates in the fact that the warming of the ocean waters off the Pacific coast of South America is usually most pronounced around Christmastime. (See glossary entries for ENSO, Jet stream, and La Niña.)

El Niño/Southern Oscillation (ENSO)

The El Niño/Southern Oscillation or ENSO phenomenon is an irregular oscillation in the climate involving interrelated changes in ocean surface temperatures and winds across the equatorial Pacific, which influences seasonal weather patterns around the world. ENSO is associated with alternations between El Niño climate events in certain years and La Niña events in others. (See glossary entries for El Niño and La Niña.)

Feedback loops

Feedback loops are cyclic flows of information through which two components of a system affect one another. Feedback loops can be either positive, through which small disturbances are amplified, or negative, in which case the disturbance is diminished. A relevant positive feedback in the climate system is the effect of temperature on the extent of ice sheets: warming reduces ice sheet size, which causes further warming because bright, reflecting ice is replaced with dark, heat absorbing rock and soil.

Fossil fuels

Fossil fuels are hydrocarbon-based energy sources formed over millions of years when the fossilized remains of dead plants and animals are exposed to heat and high pressure in Earth's crust. Fossil fuels exist in solid form as coal, shales, and methane "clathrate" (methane gas entrapped in a water-ice cage); in liquid form as oil; and in gaseous form as so-called natural gas (mostly methane). Nearly 90% of the world's primary energy production comes from the combustion of fossil fuels. When fossil fuels are burned, greenhouse gases are released into the atmosphere. (See glossary entry for Greenhouse gases.)

Fuel cell technology

A fuel cell can be used to convert fuel energy into electricity in a manner similar to, but distinct from, a battery.

In a fuel cell, electricity is generated from the reaction of chemical fuel stored at one end of the cell with an oxidant at the other end of the cell. Unlike in a conventional battery, reactants are consumed during the operation of the fuel cell and must therefore be replaced for continuous electricity production. The primary application of fuel cell technology is in the area of transportation. There are a variety of fuels that can potentially be used in fuel cells. If hydrocarbon-based fuels are used, fuel cells emit only marginally less greenhouse gases than conventional carbon-based energy sources. However, alternatives such as hydrogen cell technology, which are currently being researched, could provide a carbon-free alternative to fossil-fuel based transportation.

General circulation model (GCM)

A general circulation model (GCM) is a three-dimensional numerical model used in global climate prediction and assessment. Unlike simpler energy balance models (EBMs), GCMs can be used to solve for a variety of variables including wind patterns, air pressure, atmospheric humidity, and precipitation patterns. While the most basic GCMs model the behavior of the atmosphere alone, climate modelers often use "coupled" versions of GCMs wherein the atmosphere is allowed to interact with models of the global oceans, the major (Greenland and Antarctic) ice sheets, and terrestrial ecosystems.

Geoengineering

Any of a number of approaches to mitigate against the climatic effects of fossil-fuel burning by manipulating the Earth system in some way. Geoengineering approaches largely fall into two classes: those that are aimed at removing carbon dioxide from the point of emission or the atmosphere itself, and those that are intended to counteract the warming effect of greenhouse gas buildup by reflecting a higher proportion of incoming solar radiation back to space.

Geothermal energy

Geothermal energy is generated from the heat stored beneath Earth's surface, typically through the use of steam-driven turbines. Often water is injected into the hot subsurface of Earth to generate steam. The use of geothermal energy sources dates back to the early 20th century. Geothermal power provides less than 1% of global energy production, but it is used more consistently in certain regions such as Iceland, and California and Nevada in the United States. Potential exists for more widespread use of this renewable energy source.

Gigaton (Gt)

A gigaton (Gt) is a metric unit of mass, equal to 1 billion metric tons (tonnes), or approximately 1.1 billion U.S. tons. In the context of greenhouse gas emissions, gigatons are commonly used as units for measuring global quantities of carbon dioxide or carbon. (See glossary entry for Carbon dioxide (CO_2) equivalent.)

Global warming potential (GWP)

Global warming potential (GWP) is a measure of how much a given mass of greenhouse gas is estimated to contribute to global warming relative to the same amount of carbon dioxide (see p.28). Since GWP is a measurement of the integrated warming impact of greenhouse gas emissions, it must be calculated and stated for a specific time interval, typically one century.

Greenhouse gases (GHGs)

Greenhouse gases (GHGs) are gases in Earth's atmosphere that absorb longwave radiation including the radiation emitted from Earth's surface (i.e., terrestrial radiation). Because they absorb terrestrial radiation, these gases have a warming influence on Earth's surface (referred to as the "greenhouse effect"). Greenhouse gases exist naturally in Earth's atmosphere in the form of water vapor, carbon dioxide, methane, and other trace gases, but atmospheric concentrations of some greenhouse gases such as carbon dioxide and methane are being increased by human activity. This occurs primarily as a result of the burning of fossil fuels, but also through deforestation and agricultural practices. Certain greenhouse gases, such as the CFCs, and the surface ozone found in smog (which is distinct from the natural ozone found in the lower stratosphere), are produced exclusively by human activity.

Hadley circulation pattern/ Hadley cell

The pattern of rising moist air near the equator and sinking dry air in the subtropics is referred to as the "Hadley cell" or the "Hadley circulation" after the 18th-century English amateur scientist George Hadley who first formulated a theory about this atmospheric circulation system. The Hadley circulation is a key component of the general circulation of the atmosphere; it helps to transport heat from the equatorial region to higher latitudes and is responsible for the trade winds (easterly surface winds) in the tropics (see p.99).

Hydropower

Hydropower is power produced by capturing the kinetic energy of moving water. It is currently the most commonly used source of renewable energy, responsible for just over 6% of global energy production. Hydropower has been used in primitive forms (e.g., for powering gristmills or for irrigation) for many centuries. While obtaining direct mechnical energy from hydropower requires proximity to a moving water source, modern conversion of hydropower to electric power allows long-distance transport of energy, albeit with energy loss (that increases with distance from the source). The availability of hydropower on a regional basis in the future could be affected by shifting patterns of rainfall and runoff associated with climate change.

Ice sheets/ice age/glaciers/ glaciation/glacial/ interglacial

Glaciers are huge masses of ice formed from compacted snow. An **ice sheet** is a mass of glacier ice that covers surrounding terrain and is larger than 50,000 sq km (about 19,300 sq miles). (The only current ice sheets are in Antarctica and Greenland.) An **ice age** is a cold period resulting in an expansion of ice sheets and glaciers. This expansion is referred to as **glaciation**. Ice ages are marked by episodes of extensive glaciation alternating with episodes of relative warmth. The colder periods are called **glacials**, the warmer periods are referred to as **interglacials**.

Instrumental record

In the context of climate data, the instrumental record refers to the relatively brief record of direct measurements recorded by instruments such as thermometers, barometers, rainfall gauges, and other devices that measure atmospheric temperature, pressure, wind, humidity, and precipitation, as well as ocean temperature, salinity, water density, and currents. For the atmosphere and ocean surface only, widespread measurements are available as far back as 100 to 150 years. For the free atmosphere and deep ocean, such measurements are generally only available for the past five or six decades. A few isolated instrumental climate records from several centuries back in time are available for regions such as Europe. To obtain climate data from the distant past, climate scientists turn to climate proxy records. (See glossary entry for Climate proxy.)

Intertropical Convergence Zone (ITCZ)

The Intertropical Convergence Zone (ITCZ) is a belt of low surface pressure that is centered near the equator but migrates north and south within the tropics as the seasons change. The ITCZ is associated with trade winds that converge near the equator, ascending as warm, moisture-laden air currents that rise deep into the upper troposphere in towering cumulous clouds and rainfall-producing thunderstorms. These winds eventually sink in subtropical latitudes but not before their original high moisture content has dissipated. (See glossary entry for Hadley circulation pattern.)

Isotope

Isotopes are atoms of the same element having the same atomic number but different mass numbers. The nuclei of isotopes contain identical numbers of protons but have differing numbers of neutrons. Isotopes of a given element have the same chemical properties but somewhat different physical properties. Some isotopes are radioactive, which makes them useful for dating ancient materials (e.g., carbonaceous materials, rocks, etc.).

Jet stream

The jet stream is a high-speed wind current that lies roughly at the boundary between the troposphere and stratosphere at 8–17 km (5–11 miles) above Earth's surface. The major jet stream of each hemisphere (referred to as the polar jet stream) is located at middle/subpolar latitudes, while a weaker subtropical jet stream is found at lower, subtropical latitudes in each hemisphere. Both of the jet streams circle the globe as westerly winds (i.e., winds moving from west to east), and, like the ITCZ, shift north and south with the seasons. (See glossary entries for ITCZ, Stratosphere, and Troposphere.)

La Niña

In a La Niña event, the trade winds in the eastern and central tropical Pacific are stronger than usual, there is greater upwelling of relatively cold subsurface ocean water in the eastern Pacific, and that cold water spreads out over tropical Pacific ocean surface. During a La Niña event the tropical Pacific ocean and atmosphere are in the opposite state as they are during an El Niño event and the influence on atmospheric circulation and global weather patterns is roughly, though not precisely, the opposite. The term "La Niña" means "the girl child" in Spanish. (See glossary entries for El Niño and ENSO.)

Megaton (Mt)

A megaton (Mt) is a metric unit of mass, equal to 1 million metric tons (tonnes), or about 1.1 million U.S. tons. There are 1000 megatons in a gigaton. In the context of greenhouse gas emissions, megatons are commonly used as units for measuring global quantities of carbon dioxide or carbon. This usage should not be confused with the distinct usage of the same term to describe the explosive power of a nuclear weapon.

Metric ton

One metric ton (tonne) is 1000 kilograms, roughly equivalent in mass to the Imperial ton (2200 pounds) and 1.1 U.S. tons.

Microwave

Microwaves are electromagnetic waves with wavelengths between infrared and shortwave radio wavelengths. Microwave measurements made by satellites provide one means of monitoring atmospheric temperature changes.

North Atlantic Oscillation (NAO)

The North Atlantic Oscillation (NAO) is a measure of the strength and direction of the predominantly westerly winds that blow across the North Atlantic ocean. The measurement is based on the surface pressure difference between the subpolar and subtropical regions over the North Atlantic ocean. The size of this pressure difference, and the strength and direction of the surface winds, varies from year to year. The NAO can have a profound influence on temperature and precipitation patterns of the North Atlantic and neighboring regions of Europe and North America, particularly during winter in the northern hemisphere.

Ozone

Ozone is a compound of oxygen that is made up of three atoms (the oxygen gas we breathe contains two oxygen atoms). Ozone is an irritating gas, when encountered in surface air pollution. However, in the stratosphere, a natural layer of ozone protects life on Earth from harmful ultraviolet radiation from the Sun. CFCs have been implicated in the destruction of this ozone layer. (See glossary entries for Chlorofluorocarbons and Solar radiation.)

Paleoclimate

Paleoclimate refers to climates of Earth's past, including the early historic record of climate, to climates of billions of years ago. Methods used in paleoclimate studies (paleoclimatology) include the simulation of past climates with different continental configurations, ice sheet distributions, and greenhouse gas concentrations using climate models such as EBMs and GCMs, and the reconstruction of past climates from empirical data such as proxy climate records. (See glossary entries for Climate proxy and GCM.)

Photosynthesis

Photosynthesis is the process by which green plants and certain other organisms synthesize carbohydrates from carbon dioxide and water using light as an energy source.

The common form of photosynthesis releases oxygen as a byproduct. Photosynthesis is part of the carbon cycle and is instrumental in removing carbon dioxide from the atmosphere. (See glossary entry for Carbon cycle.)

Proxy

See Climate proxy.

Salinity

The technical term for saltiness in water is salinity. Salinity influences the types of organisms that live in a body of water and the kinds of plants that will grow on land fed by groundwater. Salt is difficult to remove from water, so salt content is an important factor in human water use. The salinity differences between different water masses in the ocean is also a key determinant of large-scale ocean currents.

Solar radiation

Solar radiation is the radiation emitted by the Sun, which generates energy from the nuclear fusion reactions in its interior. Most solar radiation is in the form of visible light and some is in the form of ultraviolet, infrared, and other wavelengths of radiation. (See glossary entry for Ozone.)

Stratosphere

The stratosphere is the layer of Earth's atmosphere that lies above the troposphere, extending from roughly 8–17 km (5–11 miles) above the surface (lower near the poles and higher near the equator) to about 50 km (31 miles) above the surface. The lower stratosphere contains a natural ozone layer, which absorbs ultraviolet solar radiation, warming the surrounding atmosphere. The warming of the atmosphere with height inhibits vertical air currents, making the stratosphere a highly stable regime of the atmosphere, in contrast to the troposphere that lies below it. (See glossary entry for Ozone.)

Sustainability

The study or application of principles aimed at creating a harmonious coexistence of human civilization with the natural environment.

Trade winds

Trade winds are the easterly winds (i.e., winds that move from east to west) that are found near Earth's surface in tropical regions. The rising atmospheric currents found within the ITCZ are associated with the convergence of trade winds. The El Niño/Southern Oscillation (ENSO) is associated with a periodic alternation between weakening and strengthening trade winds in the eastern and central tropical Pacific. (See glossary entries for ENSO and ITCZ.)

Troposphere/free troposphere

The troposphere is the lowest layer of Earth's atmosphere, extending from the surface of our planet to between 8 and 17 km (5 and 11 miles) (lower near the poles and higher near the equator). The troposphere is sometimes subdivided into a planetary boundary layer, where the atmosphere is in contact with Earth's surface, and the free troposphere, defined as the layer of the troposphere above the planetary boundary layer. The boundary between the troposphere and the stratosphere above it is called the tropopause. The troposphere contains roughly three quarters of Earth's atmosphere by mass. What we normally think of as "weather" takes place almost exclusively within the troposphere. (See glossary entry for Atmosphere.)

Index

Note: References in **bold** refer to the Glossary.

A

acid rain, 200, **214**
adaptation to climate change, 151–63
  adaptation or mitigation, 154–5, 165
  agriculture, 162–3
  economy, 156–7
  ecosystems, 152–3
  sea levels, 158–9
  vulnerabilities, 154–5
  water management, 160–1
aerosols, **214**
  geoengineering, 193
  industrial activity, 19, 47, 200
  volcanic eruptions, 18
Africa
  deforestation, 188, 189
  impacts of climate change, 77, 103, 113, 135, 139, 140, 145,147
agriculture
  greenhouse gas emissions, 14, 27, 168, 169, 184
  impacts of climate change, 117, 140–1, 162–3
  mitigation potential, 167, 184–7
agriculture sector, 167
air pollution, 136–7, 180–1
alternative energy, 170–1, 182, 212
Amazon
  rainforest, 188
  river basin, 77, 133, 145
amphibians, 130–1
Annan, Kofi, 200
Antarctica
  ice sheet, 15, 30, 104, 105, 110, 123, 148, 208
  seasons, 11
Anthropocene, 50, **214**
Arctic Ocean
  Northwest Passage, 138
  in the past, 41, 67
  permafrost, 148, 149, 208, 209
  politics, 149
  sea ice, 15, 97, 102–3, 130, 148–9
  seasons, 11
Asia
  climate, 112

Asia (*cont.*)
  deforestation, 188, 189
  impacts of climate change, 103, 139, 145, 147
Atlantic Ocean, 62, 63
atmosphere, **214**
  climate, 12–13
  composition, 12, 30–1, 32–3
  layers, 12, 38, **214**
  radioactivity, 32–3
  temperature, 38–9, 80–1
atmospheric circulation, 13
Australia, 103, 135, 145, 147, 202

B

Bangladesh, 122, 135, 206
biofuels, 170, 175, 182, 186, 207, **214**
biosphere, 68, **214**
buildings, 166, 169, 178–9, 194
buildings sector, 167

C

calcium carbonate, 108, 109, 192
carbon, social cost of (SCC), 156–7, 166
carbon capture and storage (CCS), 166, 170, 180, 181, 184, 192, 211
carbon costs, 166–7
carbon credits, 156, **214**
carbon cycle, 26–7, 184, 191, **214**
  feedbacks, 24–5, 68,106–9
carbon dioxide (CO_2), 14
  airborne fraction, 106
  changes, 41–3, 104, 105
  climate sensitivity, 84–5, 88–91
  CO_2 equivalent, 168–9, **214**
  concentrations, 31, 32, 33, 42, 43, 47, 67, 88, 84–5, 96, 116–7, 129
  direct radiative effect, 24
  emissions, 14, 27, 32, 33, 47, 106, 168–9, 170
  geologic records, 40–1
  global warming potential, 28–9, **216**
  greenhouse effect, 22
  industrial emissions, 19, 180–1
  mitigation potential, 166–7, 185
  natural fluctuation, 32–5
  possible future scenarios, 92–3
  proxy measures, 42
  sources, 27, 169
  stabilization targets, 116–7
  see also carbon cycle

carbon footprints, 196, 198–9, **214**
carbon isotopes, 32–3, **217**
carbon neutral, 186
carbon sequestration, 166, 180, 181, 184, 192, 210
carbon sinks, 188, 190, 192
CCS *see* carbon capture and storage
CFCs (chlorofluorocarbons), 12, 14, 28–9, 169, 201, **214**
China, 139, 171, 174, 175 187, 204
Clean Air Acts (1990), 201
climate, 10–11
  atmosphere, 12–13
  greenhouse effect, 14
  in the last interglacial, 66–7
  latitude, 10–11, 13
  oceans, 11
  throughout history, 14
climate change,
  geologic records, 40–1, 43
  greenhouse effect, 14, 17
  natural impacts, 18, 78–81, 87
  potential impacts, 119–49
  predictions and projections, 20–1, 50, 68–75, 83, 92–101, 117
  reasons for change, 19
  vulnerability, 154–5
  see also climate sensitivity; human impacts on climate
climate models, 17, 68–81
  climate sensitivity, 84–5, 86–7, 90-1
  complex models, 69
  Energy Balance Models (EBMs), 64, **215**
  "fingerprints", 78–81
  General Circulation Models (GCMs), 69, **215**
  predicted vs observed trends, 72–3
  regional vs global trends, 74–5
  reliability, 69
  simple models, 68
climate proxy, 42, 48, 86, 90–1, **215**
climate sensitivity, 84–91
  deep ocean temperature, 85
  definition, 84
  equilibrium climate sensitivity, (ECS), 78, 86, 94, 95
  estimations, 84
  evidence from deep time, 88–91
  evidence from past centuries, 86–7
climatic bands, 11
climatic zones, 11, 76–7, 131
  novel climates, 76–7

Image and Data Credits

The publisher would like to thank the following for their kind permission to reproduce their photographs:

(Key: a-above; b-below/bottom; c-center; f-far; l-left; r-right; t-top)

1 Corbis: Ted Soqui / Ted Soqui Photography. **2-3 Getty Images:** Michael Bocchieri. **4 Corbis:** Alberto Garcia (cra). **5 Corbis:** Rafiqur Rahman / Reuters (cb). **Getty Images:** Michele Falzone (ca); Jeff J Mitchell (tr). **NASA:** NOAA (cra). **6-7 Corbis:** Gene Blevins / ZUMA Press. **10 Science Photo Library:** Friedrich Saurer (crb). **12 Corbis:** Photolibrary (cb). **15 Corbis:** James L. Amos (b). **16 NASA:** NOAA / GSFC / Suomi NPP / VIIRS / Norman Kuring (cla). **17 Corbis:** (cl); Ashley Cooper (tl). **Getty Images:** Paulo Cunha (bl). **18 Corbis:** 2 / InterNetwork Media / Ocean. **19 Corbis:** Ashley Cooper. **22 Getty Images:** Rob Broek (b). **23 Corbis:** NASA (tr, cr, crb). **Science Photo Library:** Friedrich Saurer (cla, cl, clb). **24-25 Getty Images:** Dougal Waters. **26-27 Corbis. 28-29 Alamy Images:** Keith Bell (bl). **30-31 Alamy Images:** CharlineXia. **30 Science Photo Library:** British Antarctic Survey (tr, cra, crb). **31 Alamy Images:** Arctic Images (tl). **36-37 Getty Images:** Pete Atkinson. **37 NASA:** JPL-Caltech (tl). **38 Corbis:** Photolibrary (br). **39 NASA:** NASA Goddard Space Flight Center Image by Reto Stöckli (land surface, shallow water, clouds). Enhancements by Robert Simmon (ocean color, compositing, 3D globes, animation). Data and technical support: MODIS Land Group; MODIS Science Data Support Team; MODIS Atmosphere Group; MODIS Ocean Group Additional use: USGS EROS Data Center (topography); USGS Terrestrial Remote Sensing Flagstaff Field Center (Antarctica); Defense Meteorological Satellite Program (city lights) (cr). **40-41 IODP / TAMU:** Arito Sakaguchi. **42-43 Science Photo Library:** Power and Syred. **42 Dorling Kindersley:** The Royal Museum of Scotland, Edinburgh (cl). Hans Kerp, Forschungsstelle fuer Palaeobotanik, Westfaelische Wilhelms-Universitaet Muenster: (clb). **44 Rex Features:** Sipa Press. **46 Alamy Images:** Photos 12 (t). **48-49 Corbis:** John Miller / Robert Harding World Imagery. **51 iStockphoto.com:** TwilightShow. **53 Corbis:** 2 / Steve Lewis Stock / Ocean (b). **56-57 Alamy Images:** Dave Chapman. **58-59 Getty Images:** Paulo Cunha. **61 Courtesy of the Climate Change Institute, University of Maine, USA (see http://ClimateReanalyzer.org):** (tr). Press Association Images: AP (b). **62-63 NOAA. 64-65 Alamy Images:** The Africa Image Library. **64 Byrd Polar Research Center - OSU (The Ohio State University, USA):** (clb); Lonnie G. Thompson (bl). Corbis: Jim DeLillo / Demotix (br). Prof. em. Bruno Messerli / Geographisches Institut - Physische Geographie - Universität Bern: (crb). **66-67 Corbis:** Mike Theiss / National Geographic Creative. **70 Corbis:** Michael Reynolds / epa (cla). Sam Kittner: (clb). **71 Alamy Images:** ZUMA Press, Inc. (tl). **Rex Features:** (clb). **72-73 Corbis:** Alberto Garcia. **74-75 NASA:** NOAA / GSFC / Suomi NPP / VIIRS / Norman Kuring. **78 Science Photo Library:** Friedrich Saurer (tl). **78-79 Dreamstime.com:** Kodym. **80 Corbis:** NASA (cl, c, cr). **81 Corbis:** NASA (cl). **Dreamstime.com:** Gisli Baldursson. **82 NASA:** NOAA / GSFC / Suomi NPP / VIIRS / Norman Kuring (cla). **83 Corbis:** Ken Cedeno (bl). **Dreamstime.com:** Chriswood44 (tl). Rex Features: KPA / Zuma (cl). 85 Argo program, Japan, **JAMSTEC:** (bl). **86-87 Corbis:** Dennis di Cicco (bc). **86 The Galileo Project:** (bl). **90-91 Getty Images:** Michele Falzone. **95 NASA:** SDO (tr). **Rex Features:** KPA / Zuma (b). **96-97 Dreamstime.com:** Chriswood44. 100 **Corbis:** Reuters (bl). **102-103 NASA:** NASA Earth Observatory image by Jesse Allen and Robert Simmon, using data from the Level 1 and Atmospheres Active Distribution System (LAADS) (tr). **104-105 Corbis:** 145 / Steven Puetzer / Ocean (b). **105 NASA:** NASA images by Jesse Allen and Robert Simmon, based on MODIS data (tl, tr). **107 Science Photo Library:** Susumu Nishinaga (br). **108-109 Alamy Images:** Alex Fieldhouse. **108 Science Photo Library:** David Scharf (cb). **110-111 Robert Harding Picture Library:** Roger Braithwaite (c). **112 Getty Images:** Ron Dahlquist. **113 Corbis:** Dewitt Jones. **114 Press Association Images:** Matt Rourke / AP (t). **115 Corbis:** Ken Cedeno (cb). **118 NASA:** NOAA / GSFC / Suomi NPP / VIIRS / Norman Kuring (cla). **119 Corbis:** Chris Newbert / Minden Pictures (bl); David Turnley (bl). **Getty Images:** David McNew (cl). **120-121 Press Association Images:** Tony Gutierrez / AP. **121 NASA:** NOAA (cr). **123 Corbis:** Julie Dermansky (b). **124-125 Alamy Images:** David R. Frazier Photolibrary, Inc. **125 Alamy Images:** blickwinkel (tl). **126 Corbis:** Chris Newbert / Minden Pictures. 127 AIMS - Australian Institute of Marine Science. **128 U.S. Geological Survey:** (cla, cra). **128-129 Dorling Kindersley:** David Peart. **130 Corbis:** Olaf Krüger / imageBROKER (bl). fogdenphotos.com: (tr). **131 Corbis:** NASA (tc, ca, ca/., c, cb, bc). **132 Science Photo Library:** Jim Edds. 133 Corbis: Paulo Whitaker / Reuters. **134 Alamy Images:** Robert Harding Picture Library Ltd (tr). **Getty Images:** Andy Ryan (bl). **135 Corbis:** Ashley Cooper (tl); Rafiqur Rahman / Reuters (bl); David Turnley (tr). **136-137 Getty Images:** David McNew. **138-139 Getty Images:** Ira Block. **141 Corbis:** David Turnley (bl). **144-145 Alamy Images:** US Army Photo. **146-147 Getty Images:** Matt Cardy. **148-149 Alamy Images:** Accent Alaska.com. **150 NASA:** NOAA / GSFC / Suomi NPP / VIIRS / Norman Kuring (cla). **151 Alamy Images:** Purestock (cl). Corbis: Robert Landau (bl). Getty Images: Jeff J Mitchell (tl). **152-153 Sun-Sentinel:** South Florida Sun-Sentinel. **154-155 Corbis:** Imageplus. **156 Corbis:** Robert Landau (cr). **158 Getty Images:** Jeff J Mitchell. **159 Alamy Images:** Hoberman Collection (tc); Joern Sackermann (tl); Lancashire Images (tr). **160-161 Dreamstime.com:** Instock. **162-163 Corbis:** Lu Wenxiang / Xinhua Press. **163 NASA:** NOAA (l). **164 NASA:** NOAA / GSFC / Suomi NPP / VIIRS / Norman Kuring (cla). **165 Corbis:** Abir Abdullah / epa (bl); 237 / Ron Bambridge / Ocean (tl). **Dreamstime.com:** Simon Thomas (cl). **166-167 Corbis:** Frank Lukasseck. **166 Alamy Images:** BonkersAboutPictures (clb). Corbis: Imaginechina (crb, fcrb). **167 Corbis:** Alan Schein Photography (fcrb); Imaginechina (fclb); Marc Dozier (clb); Everett Kennedy Brown / epa (crb). **168-169 Corbis:** Sebastian Kennerknecht / Minden Pictures. **170-171 Alamy Images:** Darrin Jenkins. **170 Corbis:** US Air Force - digital version c / Science Faction (clb). **172-173 Corbis:** 237 / Ron Bambridge / Ocean. **172 Corbis:** Shuli Hallak (tr). **175 Alamy Images:** FLPA (tl). **176-177 Dreamstime.com:** Simon Thomas. **178 Corbis:** Jose Fuste Raga (l). **179 Corbis:** David Denoma / Reuters (b). **180-181 Statoil:** Bitmap / Kim Laland for StatoilHydro. **182-183 Corbis:** Topic Photo Agency. **184-185 Corbis:** Markus Hanke / www.MarkusHanke.de. **186-187 Dreamstime.com:** Jochenschneider. **188-189 Corbis:** Ton Koene, Inc / Visuals Unlimited. **190 Corbis:** Mark Thiessen / National Geographic Creative (b). **192-193 John MacNeill Illustration. 194 Corbis:** Mischa Keijser / Cultura (bl); Tetra Images (bc). **Dreamstime.com:** Haiyin (cb). **195 Alamy Images:** D. Hurst (cb, br); Janusz Wrobel (crb). **Dreamstime.com:** Pixattitude (bl). **Getty Images:** Rubberball / Mike Kemp (t). **196 Pennsylvania State University:** (crb). **197 Alamy Images:** National Geographic Image Collection (tl). Pennsylvania

State University: (b). **200 Science Photo Library:** Will & Deni Mcintyre. **201 Alamy Images:** Visions of America, LLC (bl). **202 Getty Images:** Yvan Cohen (tl). **203 Getty Images:** Cris Bouronocle (br). **204-205 NASA:** NOAA / GSFC / Suomi NPP / VIIRS / Norman Kuring. **206-207 Corbis:** Abir Abdullah / epa. **208 Corbis:** Ashley Cooper / Picimpact. **209 Getty Images:** Karen Doody / Stocktrek Images. **210-211 Corbis:** Steven Kazlowski / Science Faction. **212-213 NASA:** NOAA

Cover images: *Front:* **Alamy Images**: Design Pics Inc crb; **Getty Images**: Mint Images - Frans.Lanting ca; Back: **Corbis**: Talia Herman crb, John Hyde/Design Pics fcrb.

All other images © Dorling Kindersley
For further information see: **www.dkimages.com**

The publisher would like to acknowledge the following sources of information used in this book:

p.34 graph of global average surface temperatures, based on data from NASA. **p.37** graph of ocean heat content, based on data from Nuccitelli et al (2012) (www.sciencedirect.com/science/article/pii/S0375960112010389). **p.45** graph of oxygen content of the Mar Sci (2010). **p.50** graph of long-term global temperatures based on a chart by Jos Hagelaars, adapted by David Spratt, in www.climatecodered.org/2014/05/the-real-budgetary-emergency-burnable.html. **p.52** graph of dry area percentage in the western U.S. based on data from http://www1.ncdc.noaa.gov/pub/data/cmb/sotc/drought2014/06/Reg120_wet-dry_bar01000614.gif. **pp.52–53** map of pattern of U.S. drought in August 2014 based on a map from http://droughtmonitor.unl.edu/data/pdfs/20140805_usdm.pdf. **p.56** graph of hottest year on record

based on data from NOAA. **p.57** graph of temperatures in St. Louis based on information from U.S. National Assessment. **p.60** graph of frigid nights in Peoria based on a graph from Climate Central (original courtesy of Heidi Cullen and Dennis Adams-Smith). **p.61** global surface temperature map graphic based on information from The Climate Reanalyzer project http://cci-reanalyzer.org. **p.65** map of Kilimanjaro based on information from http://www.nsf.gov/news/mmg/media/images/icemap_h.jp. **p.76** maps of disappearing and novel climates based on information from Williams et al article: http://www.pnas.org/content/104/14/5738.ful.pdf+html. **p.91** estimates of climate sensitivity graphic based on a graphic adapted by Michael Mann from his 2014 article in Scientific American. **p.94** graph of the faux pause based on a graphic by Michael Mann from a 2014 article in Scientific American. **p.97** graph of sea ice extent based on data from Copenhagen Diagnosis (2009) and NSIDC: http://nsidc.org/arcticseaicenews/files/2014/10/monthly_ice_NH_09.png. **p.176** graph of growth of electric vehicles based on data from http://upload.wikimedia.org/wikipedia/commons/f/f6/US_PEV_Sales_2010_2013.png. **p.182** graph of gallons of water used to produce 1 million BTUs based on information from the World Policy Institute-EBG Capital analysis based on U.S. Department of Energy 2006 and World Economic Forum and Cambridge Energy Research Associates, 2009. **p.182** graph of water consumption of an 18,000 BTU air conditioner over a week based on information from the World Policy Institute-EBG Capital analysis based on U.S. Department of Energy 2006, World Economic Forum and Cambridge Energy Research Associates, 2009, and http://www.consumerenergycenter.org/home/heating_cooling/window_ac.html. **p.189** map of rate of change in forested area based on the map in MC Hansen et al. Science 2013;342:850–853.

Acknowledgements

The authors would like to thank their colleagues at Penn State, who have helped to foster a stimulating environment for discourse on the important topic of climate change and its impacts. These include, among many others, Richard Alley, Mike Arthur, Eric Barron, Tim Bralower, Sue Brantley, Bill Brune, Rob Crane, Ken Davis, Bill Easterling, Jenni Evans, Bill Frank, Kate Freeman, Sukyoung Lee, David Pollard, Jim Kasting, Klaus Keller, Ray Najjar, Art Small, Petra Tschakert, Anne Thompson, and Nancy Tuana. Michael Mann would like to acknowledge his colleagues at RealClimate.org for the many keen insights into the science of climate change that they have shared over the years. Lee Kump similarly acknowledges his fellow members of the Canadian Institute for Advanced Research.

We would also like to thank our colleagues who provided helpful comments on the book at its various stages, including Richard Alley, Klaus Keller, and Jean-Pascal van Ypersele. Our thanks to the following reviewers who supplied helpful feedback as we prepared the second edition: Jason Allard (Valdosta State University), Dawna Cerney (Youngstown State University), Marla Conti (Glendale Community College), Mike DeVivo (Grand Rapids Community College), Danielle DuCharme (Waubonsee Community College), Michael Farrell (Los Angeles City College), Larissa Hinz (Eastern Illinois University).

We also extend out thanks to Christian Botting, Crissy Dudonis, and Karen Gulliver at Pearson, and the staff at Dorling Kindersley for their superb editing and design skills, which benefited the book greatly, both in its previous edition and this current, new edition.

Michael Mann dedicates this book to the memory of his brother, Jonathan, and to his daughter, Megan, who will grow up on an Earth whose destiny rests in our hands. He thanks his wife, Lorraine Santy, for her support, and his parents, Larry and Paula Mann for the encouragement they have always provided. Lee Kump dedicates this book to the newest member of his family, his son-in-law Matt Kohler, who distinguished himself during two tours of duty as a Marine in Iraq and Afghanistan.

Dorling Kindersley would like to thank: Deepak Negi for picture research; Ed Merritt for cartographic work; Helen Spencer for initial design work; Miezan van Zyl, Wendy Horobin, and Helen Fewster for editorial assistance; Parnika Bagla and Hina Jain for indexing; Anjana Nair and Shreya Anand for their work on graphs; Jaypal Chauhan for technical assistance; and Starr Baer for proof-reading. The following people worked on the jacket: Dhirendra Singh (jacket designer), Harish Aggarwal (senior DTP designer), Saloni Singh (managing jackets editor), Claire Gell (jackets editor), and Sophia MTT (jacket design development manager).

Dorling Kindersley would also like to thank the following people for their work on the first edition: Sophie Mitchell; Richard Czapnik and Stuart Jackman for design work; Steve Setford for editorial work; Ed Merritt for cartography; and Silvia La Greca Bertachhi for production. Additional thanks are due to Clive Savage, David McDonald, and Johnny Pau for design work; Sue Malyan and Jenny Finch for editorial work; and Sue Lightfoot for the index.